Crafting Frames
Of
Timber

Also by Michael Beaudry:
The Axe Wielder's Handbook

Crafting Frames
Of
Timber

Michael Beaudry

Mud Pond Hewing and Framing
Montville, Maine

Copyright 2009 by Michael Beaudry

All Rights Reserved.

No part of this book may be reproduced in any form whatsoever without prior written permission of the author, except in the case of brief passages embodied in critical reviews and articles

ISBN# 978-0-578-01837-9

Published by:

Mud Pond Hewing and Framing
40 Peavey Town Road
Montville, ME 04941

www.mudpond.net

To my loving wife, Claudette, who has had to dwell in my architectural experiments, not all of which have been successes.

And to my brother, Den, who has shared the passion for hewing with the Broad Axe and building frames of hewn timber.

Special Thanks go to the following:

Aaron Sturgis and the Preservation Timber Framing Crew, Den Beaudry, Bill Behrens, George Fuller, Robert Hitt, Rick Irons, Jim Lattin, Paul Martin, Bud Menard, John Phelan, Bill Rispolli, and Stan Steinhoff for enriching my field of framing experience.

Mary Ellen Hitt, Jim and Harmony Lattin, Lillian Martin, Greg Morley, Claudette Nadeau, Connie O'Brient, Al Oliver, Bill Rispoli, Anne Sheble, and Penny West for photos.

Contents:

Hand Crafting Frames	pages 9-15
Planning	pages 17-32
Trees	pages 33-49
Hewing with the Broad Axe	pages 51-82
Tools	pages 83-95
Mortise and Tenon	pages 97-109
Sills and Floor Joist Systems	pages 111-128
Bracing	pages 129-142
Tying it all together	pages 143-161
Roof Systems	pages 163-193
Cruck Framing	pages 195-203
Scarf Joints	pages 205-221
Shingles and Trunnels	pages 223-234
Raising Day	pages 235-247
Building with Logs	pages 249-266
In the End	pages 267-271
Further Reading	pages 273-274
Index	pages 275-277

8

Hand Crafting Frames

Every hand crafted shelter, built with local materials, is a story. It stands in sharp contrast to the pre-fabricated houses all too often being built today. There is an intimacy in the designing of the house, an intimacy in working with the local materials, and an intimacy working with the tools. The product, regardless of the skills of the builder, stands unique.

I live in Maine. It is perhaps the most forested state in the nation. The percentage of land covered in forests now approaches what it was when the Pilgrims first landed. It is a sparsely populated state when compared to other states. Despite all this, the bulk of the building materials used in Maine are imported. Few people are involved with building their own structures. The hand-crafter stands alone against this wave. The materials chosen have a natural semblance of place.

I have been involved with crafting shelter for more than 30 years. The early years were crude manifestations of my limited skill. With time, however, we learn and the structures we leave behind display a deepening understanding of craftsmanship. Most of those years I did not build professionally. I generally built structures upon my farmstead, or worked with friends putting up timber frames upon their land. Only in the latter third of those years have I ventured out as a professional builder. Even as a professional, I have managed to continue the handcrafted uniqueness of building by continuing to work directly with the owners. Sometimes they merely design and locate the structure, but all too often they roll up their sleeves, grab a mallet and chisel and go at it, cutting joinery and fitting timbers right along with me. The level of the owner's involvement becomes, frequently, a measure of the satisfaction I feel with the job. It becomes a shared experience rather than merely a contracted arrangement.

I recall back in 1980, cutting joinery for a barn we were building out of recently hewn timbers, when Bill, the future owner of the barn, broke away from his cutting and said "This is how I want to live my life." That succinct phrase said it all. It described the creative joy we were experiencing as well as making it clear there was a philosophical understanding that made sense of what we were doing.

Philosophy plays a major role in hand crafting shelter. I believe philosophy is one of the major factors determining handcrafted buildings.

Aesthetics also plays high in hand crafting. We may want our building to approach some artistic form that we have imagined or designed. The uniqueness of this form may require skills beyond those that we currently possess, requiring us to reach deep within ourselves to find, learn and develop these abilities. Finding the proper materials and working them into an expression of artistic form can add considerable time and expense to a project.

Unfortunately economics plays a role. The costs may eventually rise above what was expected and the struggle with the miscalculated economics affects the designs of our lofty aspirations. Compromises must be made.

I have always placed philosophy first. Claudette and I have chosen to live close to the land: growing our own food, raising animals, heating our home with wood, building our home from trees upon the land. This, almost subsistent, view of the "good life" does not promote economic abundance.

Philosophy dictates that the buildings I put up should be of materials from the land worked by hand. Aesthetics envisions a highly skilled crafted structure with a very natural appearance sitting harmoniously with the surroundings. Economics tears into this vision and assists it. There is neither money nor time to go and acquire the skills through some workshop or seminar, so the skills have to be acquired on the job as I go. Limited available money forces compromises both in form and size. There is a limited amount of time. Winter comes too fast and either the project sits idle or additional compromises are made to get it finished.

Deriving building materials directly from the land with hand labor has been the philosophy that has impacted my building decisions.

If one discards the time invested, there is no cost for me to walk into the woods, fell a tree and hew it into a dimensional timber. Pulling away from the cash economy, encourages one to produce one's own building materials, whether that be building poles, timbers, shingles, etc. This is totally in harmony with my "good life" philosophy. It also adds to the aesthetic pleasure I experience both in producing my own building materials as well as living in a building of these hand shaped materials.

If one has a Broad axe, one can create ones own timbers.

My particular passion lies with the bones of structures. Felling trees and hewing out dimensional timbers, cutting the joinery, and raising a timber frame or log shell gives me the greatest satisfaction. There is a profound sense of accomplishment when one successfully hews a log into a square beam, or successfully cuts a tight fitting joint. The crowning point for me, however, is the finished frame. All of the work cutting and joining is revealed as the frame stands tall and golden against a setting sun on raising day.

Once while in the process of raising a frame, a person assisting was feeling fatigued from all the climbing and lifting that was expected. She turned to me and asked, "Do you do this every day!" Well the answer is yes and no. It depends upon what one means by *this*. If she was asking do I raise frames every day, the answer is clearly "no." Raising the frame can take one, two, maybe three days depending upon the size of the frame and the number of hands assisting in the raising. The processes of collecting the materials, hewing the timbers, cutting the joinery add weeks or even months to the process. It is this process that makes the handcrafted structure a story. This process can diverge radically from conventional building, requiring alternative ways of even thinking about building. And when we venture into uncharted waters, we can encounter problems that mainstream thinking has no solutions for. We must think, act, and improvise.

I was never very good at picture taking in the beginning. I never felt they really added anything to the experience. As I felt a need to develop a portfolio, it was pictures of finished frames that I believed I needed. People, however, that I worked for or with would give me pictures of the process as it progressed. Eventually I saw the value in these pictures. The process is the story. As Robert Hitt and I were creating the cruck frame (depicted above), his daughter, Mary Ellen, was documenting it in photographs. Once the frame was raised she gave me a copy of the CD she had created. In it she had captured intimate details easily lost. She had captured techniques in scribe fitting. She had also captured the joy of building, as well as the frustration as winter closed in upon our efforts. She saw and captured the story. This enabled me to rethink my own past, to look at each building as an adventure, as a story.

As we approach hand-crafting structures we need to consider possibilities. We need a little introspection. We have to come to rest with our basic philosophy of life. The building, whatever shape it is to take, must be in harmony with that philosophy. It must be aesthetically pleasing. Why go through the effort if it isn't. One can easily import a used trailer to the site if neither philosophy nor aesthetics dictate otherwise. Lastly, it must be realistic. Realistic is simply another term for economically feasible.

If I had a recommendation it would be to start each project on the inspirational side of the bell curve, not the practical. Do not look at specifics. Read and look at pictures. Absorb ideas. Immerse oneself in ideas. When one strikes a resonant chord within, pursue it. Move from inspiration and focus totally upon the techniques and practices that will make the idea a concrete reality.

With each building we begin the process anew. Even if one has built thirty buildings, the next one can and should be a new experience. Study what new ideas are out there. Allow these ideas to mingle with ideas you already possess. If something strikes you, pursue it. Make it your own.

Recreating a German Pioneer Cabin

Because my passion lies with the bones of a building, this work is concerned with the anatomy of timber frames and log shells. Because each building has its own story, this book is concerned with variations inherent in the buildings themselves and the techniques necessary to create these structures.

I have limited this work only to projects that I have personally been involved in, enabling me to speak from direct experience rather than simply from some intellectual understanding. Though I have attempted to cover many possible variations, one must bear in mind these but scratch the surface of what is possible.

Buildings, even small buildings, are major investments in time, money and effort. Buildings are large, heavy and not easily moved.

Hand crafting a building does require taking an honest look at oneself to see if it really is in one's nature to tackle such a project.

I know a man who was going to build a cob house. He had read the literature. He had experimented with the mix and had come up with a combination of ingredients that would have created a beautiful house.

He had skipped perhaps the most important ingredient, his own reality. In the past he had become a master of the ultra-light shelter.

He was building quick, easily heated, shelters of no more than 2"x 2" studs with half-inch boards and translucent fiberglass panels. The structures formed parabolic curves of amazing strength and beauty. They could be erected quickly. They were small, therefore, a small woodstove was all that was needed to heat them.

Cob is a labor-intensive material to work with. For some, this would be the experience of a lifetime. For this individual, it made no sense. No progress was made on the rubble foundation. No cob walls were ever made. No poles for the frame were ever cut or shaped. While he was waiting to build this cob structure, several *temporary* buildings were erected, each with a specific function, each well built and able to withstand the Maine winter.

He envisioned one story but he lived another. His parabolic structures have evoked awe in many. They form serviceable, beautiful buildings with so little in materials

When I look at my own philosophy, I have remained constant. When I build for others, I allow their philosophies to shape the experience. They create their own stories. Most have happy endings.

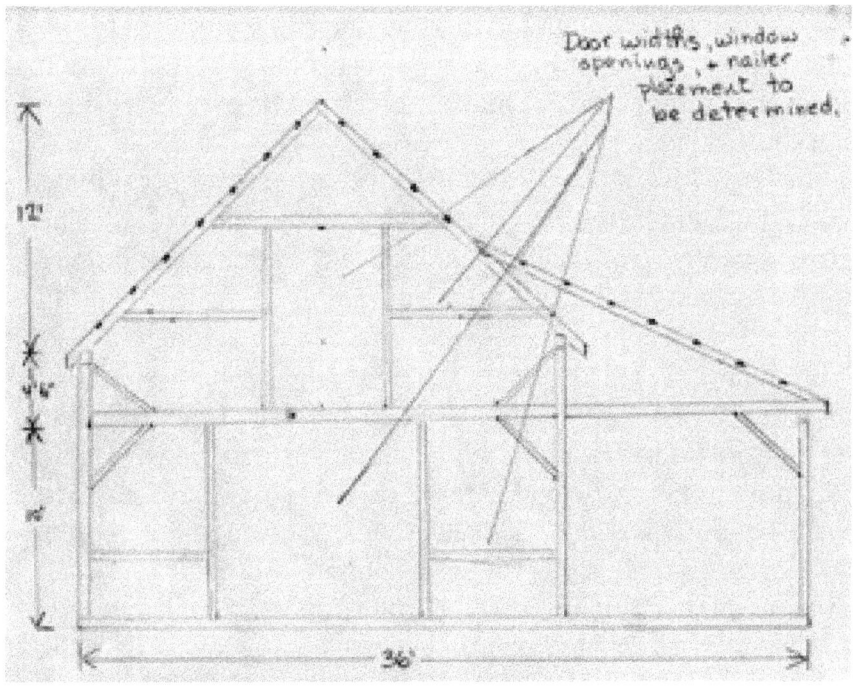

Planning

    Before any work can begin, a plan must be made. This plan may involve several parties. If permits are required, the town's code enforcement officer and the planning board may be involved. In some rural communities the town's involvement may be minimal. In many Maine rural communities, the town will issue the permit and require the permit holder to have a proper waste water system installed and inspected before any plumbing is done inside the structure. If the town has no electrical inspector, at the very least the public utilities company will do an inspection of the service head, the meter and the circuit breaker box before hooking electricity to the house.
    If the proposed structure is waterfront property or sitting on a resource protection zone, setbacks and other requirements may be required that will greatly affect the planning process.
    If a bank is to loan money for the project or an insurance company will be expected to insure the finished structure, they will want to know the plan and may require that aspects of the plan be modified or revised.

In addition to the above, anybody brought on to work on a specific aspect of the project will need a plan: the foundation contractor, the framer, roofer, mason, electrician, plumber, finish carpenter etc.

One can hire an architect or an engineer. In designing and building a handcrafted structure this is seldom the case. The transition, however, from having an architect design a building to designing it oneself is a large step that some of us fail to fully understand.

Frequently I am asked if I would like to help someone build a frame. I ask what they intend to build and the response may be, "I want to build a 20' x 30' shop." I then ask if it is to be a one story, one and a half story, or two-story building. In response I get a blank stare. I usually design frames for people, giving them a general timber list, a sketch of the frame and, sometimes, general illustrations of the joinery I will use. I cannot, however, even begin until I get a basic sketch of what they want.

I was involved in cutting five large frames for the Maine Department of Inland Fisheries and Wildlife. The department's engineer supplied each of us working on the frames with clear blueprints. The blueprints showed the overall frame, the lengths and dimensions of the timbers along with clear illustrations of the joinery to be cut on each timber. Though all five frames were different widths and lengths, the basic design was kept essentially unchanged. It was a bit like factory work.

When we had finished cutting joinery, all the timbers were transported to the site and several of us, along with two cranes, a man-lift and other pieces of heavy equipment, raised the frame. None of the pieces were test-fitted prior to the raising. As long as each joiner measured and cut to the exact specifications of the blueprints, the frames rose just as planned.

Such a structure, however, is very linear. Most people wanting a hand-crafted frame are not engineers. They generally have an artistic flair and want this artistry to be incorporated into the design.

Frame for Maine Dept of Inland Fisheries and Wildlife being raised.

I recently had the opportunity to work with the finest planner I have ever met, Rober Hitt.

Robert Hitt

I had known Robert for years. I was working on a community project that involved hewing and framing. He called me and expressed an interest. I told him we meet once a week on this volunteer project and that if he wanted to join us he was free to do so and I would be happy to instruct him. He did, and on our free time he began talking about a house he wanted to build. He said he always built or renovated out of necessity but never really built something that he personally designed. He wanted it to be different with the natural curves of the wood being prominent. He also wanted it to be fairly small but complex. He asked me if I would be interested in working with him on the project. My ears always perk up when someone mentions building something unique. I introduced him to the concept of the cruck frame. He invited me over to his property and together we walked the woods looking for oaks that would make natural crucks.

He then drafted up some design plans involving 6 bents with a shed addition and a couple of dormers. I took the plans home, looked at how the structure could be joined. I found a few problems, brought my concerns back to him. He did some further revising and again gave me plans. In the back and forth dialogue, I slowly came to understand what he wanted and he became clear on how the structure must be joined.

In the course of designing he learned about cruck framing, English tying joinery, purlins, rafters, the placement of braces, and plank wall construction. He thought through the issues of ceiling heights, stairwell and chimney placement, windows, doors, insulation, and roofing.

The result was that I was given a complete set of plans that showed heights, widths, lengths of every part of the frame. The drawings showed the crucks, curved braces, flared posts. The sizes of the timbers were clearly listed. I told him with these plans there would be no problem. I could easily engineer the frame.

He involved the foundation person in the same way, showing what he wanted, taking the man's suggestions and modifying and improving his design.

Robert, with drawings completed, wanted to see the plans in three dimensions. Using balsa wood, he created models: one of the finished house, one of the frame and one of the foundation.

Model of complex foundation

Model of cruck frame

It was not merely the house that his planning focused upon. It was the setting. He would frequently climb to the top of the field where he envisioned the house. There was a beautiful view at the top of the field, but with his artistic vision he saw that if the house was recessed a little further up the hill into the woods, and an opening created in which the view was framed by the woods, the view appeared more dramatic.

The driveway skirted the field, entered the woods and approached the house from behind. It was not visible at all from the front of the house. Only the field in front, and the lake and hills in the distance could be seen from the house.

He quickly saw that in his planning he needed early on to make contact with a competent woodsman who could share his vision. That man was Moe Martin. Moe arrived with skidder, tractor, loader and portable saw mill. Moe cut and stumped the winding path that would become the driveway. He cut the channel for the underground power. He cleared the house site. All sawable timber was stacked by the mill. Firewood and pulp were stacked in 4' lengths near the proposed driveway.

Once driveway, foundation and septic system were in place, I was called. The decision had previously been made that even the sills and floor joist system would be timber framed and that as many building materials as possible would be harvested from the property. The size of Robert's woodlot was substantial and the diversity of trees upon it was simply amazing.

Sawing out sills and joists from black locust logs.

Robert decided that he would act as his own forester and work directly with Moe in the harvest of his trees. In a lower field there was a dense stand of mature Black Locust trees. No wood is superior to Locust for both strength and rot resistance.

Robert decided that all the sills and first floor joists would be Locust and if the wood proved attractive enough, it would be used elsewhere in the frame as well.

I gave Robert and Moe the timber list for the floor system. Soon Robert and Moe were harvesting and sawing Locust beams. The timbers were then brought directly over to me while I commenced cutting the sill joinery.

Cutting sills and joists joinery.

The finished black locust floor system

Large old pasture pines felled during some of the earlier clearing provided the 1 ½" planking for the subfloor.

With that accomplished, a timber list for the entire frame was provided. The initial focus was on the timbers necessary for the basic raising with a crane.

Robert made it clear that in keeping with his original vision, all timbers in the frame would have at least one live edge, many would have two. He would be directly involved with the harvest and select the species he wanted for distinct aspects of the frame. The timbers would be sawn, the joinery cut, and the live edges peeled with drawknives. Following this, the sawn surfaces would be planned, sanded and oiled. They would then be stored under cover until needed for assembly.

I developed a code system and all timbers were clearly marked with a code letter and number so there could never be any question where in the frame the timber was to be used. A code system is essentially a grid placed upon the frame much the same way map grids are used. The bents are each given a number and the post are each given a letter.

A post labeled C4 would be the end post of the fourth bent. A tie beam labeled on one end A3 and on the other B3 would be a tie beam that spans the third bent from A to B.

This was an intricate frame with heights, angles, and lengths always changing. I kept the plans with me at all times, I used a carpenter's calculator frequently, and I kept a notebook in which I entered all sorts of information. If I made a special note when cutting a particular post, I could reference it when cutting the tie beam or plate that was to sit upon that post. Any calculations were kept in the notebook and it was used frequently.

This is perhaps the most involved frame I have ever cut. It is also one of the most labor intense. Timbers were handled repeatedly. Timbers had to be moved from the saw. The joinery had to be cut. The timbers were than peeled, loaded onto a trailer and transported to a shop where they were sanded and oiled.

Robert brought on a fourth man, Pat. Pat was exceptional with the electric hand planer, a disc sander and a belt sander. He was responsible for giving each timber its polished appearance. Once the timbers were oiled, we had to handle the timbers again, moving them into storage away from the weather, only to have to move them once more when it was time to assemble.

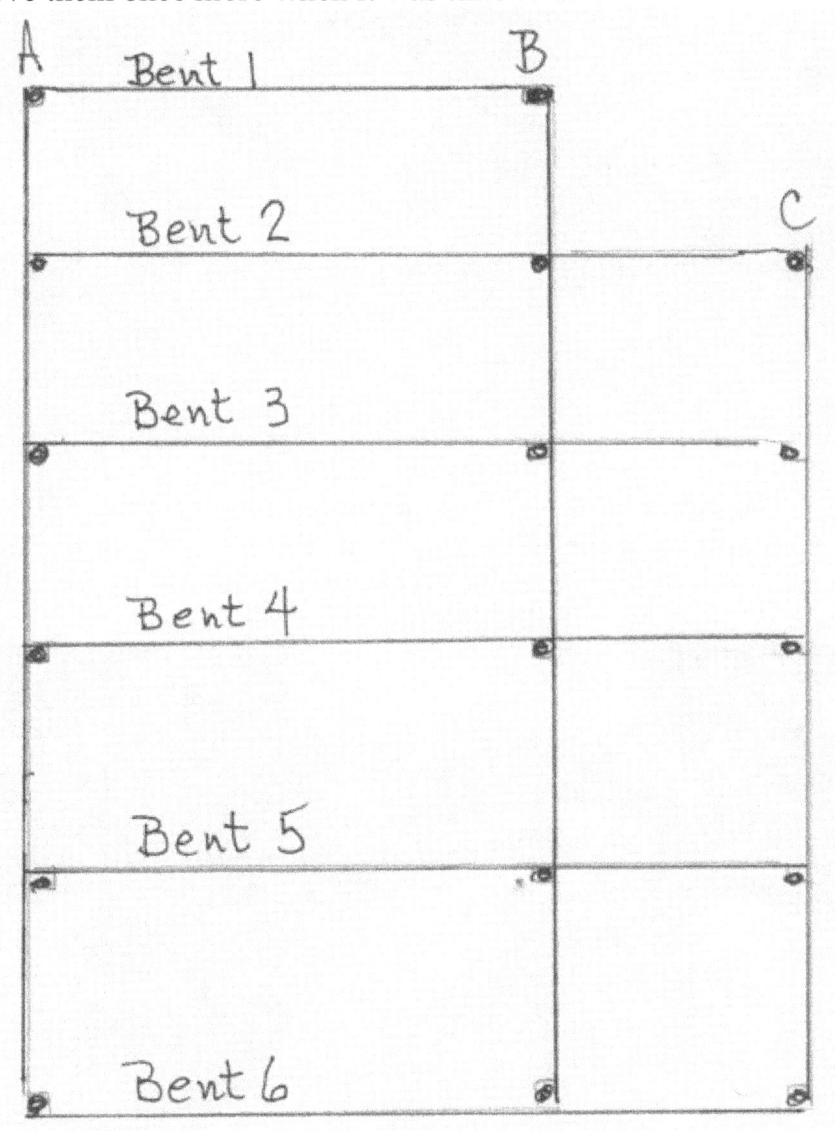

A simple code system for timber frames

Oiling timbers for the Frame.

The oak crucks were too heavy to move around. Using heavy equipment these were moved to the deck. Once on the deck it took pry bars and rollers to maneuver the heavy oak timbers to where they could be scribe fitted and pegged.

Pry bars and rollers were necessary to maneuver the oak crucks.

Taking a break while assembling cruck pairs.

All three crucks were assembled one on top of the other. Once assembled the crucks had to be lifted by crane and stored until the raising.

For Robert, the planning and coordinating had been a continuous and demanding full time job, financed by equity from his property and the sale of a small parcel of land.

The crucks assembled, moved and awaiting raising.

I started cutting joinery for him in October. During the coldest part of winter the timbers were sanded and oiled. Work ceased while we waited for winter to break in late April so that the crucks could be assembled. We set a May date for the raising with the crane.

We discussed the raising. Robert and I agreed to mutually coordinate the raising. It was believed that portions should be hand raised prior to the crane day. All the post and plates were set by hand along with the shed tie beams and joists.

Post and plates were raised by hand.

The raising day was set and announced. Food was provided and many friends invited. The crane operator arrived and, in the course of the day, the crucks, remaining joists, rafters, and purlins were raised and pegged.

I believe it was the first time in a year that Robert could actually relax and enjoy all his work. Even here, however, the respite was short lived. We took one day off and we were back at it again.

Robert's plan called for an extensive wind brace system of snaky wavy trees. It also called for some very curved tension braces. In addition to these two items, a dining space that jutted from the building still had to be framed along with the two shed dormers.

Crucks raised into place.

Frame at day's end.

I provided Robert and Moe with the final timber list. Robert took total responsibility for finding these timbers. Though necessary for bracing and stability, Robert was emphatic that these timbers had to have strong aesthetic value.

He chose the trees for felling, instructed and assisted the sawyer in sawing, and planned out the simplest details in the framing so that the joinery could be engineered and cut. It was mid-July by the time the frame was complete.

The finished frame.

The frame still needed to be planked, roofed, insulated and shingled. A chimney still needed to be built. I worked with Robert planking the building and insulating and sheathing the roof. I moved on after that.

Before winter set in Robert had the building closed in and roofed. He has decided to take the coldest, snowiest part of winter off and he would resume his planning and coordinating and building in the Spring. There are still windows to put in, French doors to put in, etc. There is still wiring, plumbing, and putting up wallboard, etc. to be done. But through it all I am sure Robert will be involved and what he will accomplish in finish carpentry will be exciting and unique from any house I have seen. All who see the structure ask me about the house. They all want to see how it is finished. I, however, have no doubt about how it will be finished. It will be finished as it was started: as a work of art.

Perhaps, for many, Robert's slow and continuous planning is more than they can bare. I give him as an example.

What he is trying to accomplish is far more than most of us try to do. His dedication and constant attention to the project are what really makes a hand-crafted structure a success.

Shows roof being insulated and Sheathed. Chimney is built. Only the dormer is trimmed with rake and fascia. No windows or doors have been placed as of this photo.

He was always willing to embellish his frames without straying from his original vision. For a framer such as myself, Robert always had me thinking beyond what I knew. He always had me reaching deep inside my resources to create the home he wanted. Many people leave me feeling that we could have done more with their structure. With Robert, I felt I was always challenged and forced to work to the maximum of my ability.

As I walked away from his structure and began work with others I was still amazed at how well he thought out his frame and how enjoyable it was to work with him. He never strayed from his vision. How rare that is.

Trees

So much about hand crafting shelter revolves around the ability to shape and work wood. There are so many factors to consider when working even a single species of wood. There are grain, knots, moisture, stresses, spiral growth, sapwood, heartwood, shrinkage, checking, decay resistance etc. When one adds to this the amazing variety of species, each with its own properties, one can design and join structural members to meet virtually every need.

The woods represent real wealth. If one cultivates the skills necessary to convert trees into hand crafted shelter, the woods can provide an honest and decent livelihood, as well as providing a home.

One, however, may be blessed with neither an abundance of forest nor a diversity of tree species. Even when that is the case, people have found ways to meet their needs for shelter using wood considered inferior and/or damaged.

I can provide a story of my own to illustrate this point. In January of 1998 we were struck with a very slow moving, highly destructive ice storm. For the better part of a week the woods were alive with the almost constant explosive sound of tree trunks snapping and collapsing.

Ice Storm of '98

In its wake I was left with many broken trees. Many were simply a snapped tree trunk rising twenty or thirty feet up with neither branches nor crown.

I was devastated. It appeared as though the woods would never recover. When spring came I thought about doing a salvage operation within my pine lot. Financial and time constraints, however, kept me from doing much of anything other than clearing up the debris upon the woodland floor.

As summer progressed, the pine borers laid their eggs under the bark of the dead and standing trunks. Another winter and summer passed. By winter, two years after the storm, the bark was falling off of the standing trunks and it appeared the wood was useless for anything other than firewood.

I went out into the snow and felled one of these dead and crownless trees. I looked at it as it lay across the snow and from all appearances the wood seemed sound. Instead of cutting it into firewood, I went to the house and came back with my hewing axe. Once hewn, I was amazed at how fresh and sound the wood was. It did have a few pine borer holes, but beyond that there appeared to be nothing wrong with it.

Ice Storm of '98

Salvaged timber from standing deadwood two years after ice storm.

What soon followed was a belated salvage operation. One dead tree after another was felled. A check was made of its condition, and if sound it was hewn.

A sizeable collection of timbers was created. The salvage operation gave me enough timbers to raise both a hewn log workshop for myself and a small timber frame barn for my goats

Hewn log workshop built of salvaged pine logs.

A couple of years ago Ed Wynn, approached me and asked if I could cut a barn frame for him. He told me that he hoped to have the barn timbers sawn directly from the property. What he wanted was a long and narrow barn: only 16' wide, but 48' long. I told him I would develop a plan and a timber list for him. He asked that I come over and see the woodlot.

The woods were nothing more than an abandoned pasture that had grown into a stand of predominantly Quaking Aspen, *popple* as it is called locally. Aspen is a pioneer species that takes advantage of the open sunlight an abandoned field provides. It grows fast, is short-lived, and prone to decay. A limb will break off and decay will work its way from this break down into the heart of the tree, rendering the tree useless for anything but firewood. And in these parts, few people want to deal with it even as firewood.

Small barn built of hewn salvaged dead pine and poplar trees.

The aspens stood where he wanted the barn and barnyard, so they would have to be cut anyway. If they had to be cut and if they could be sawn into timbers, it made sense to use them.

Basically the decision was made to purchase the sills and floor joist from a mill that could supply these in Hemlock. The rest of the timbers would be popple. I told him if the core of the tree was brown instead of white, fungus and decay had entered the tree and such logs should not be sawn. If, however, the trunk was white and clear through the core, the tree could be sawn and the timber used.

He felled the poplars and yarded them up. He hired a man with a portable sawmill, who came to the site, and using the timber list I provided, sawed out all the timbers needed to erect the frame.

The barn was set on rubble piers and in the winter of 2006/2007 the barn frame was cut and raised. Ed closed the barn in with Pine also sawn by portable mill.

Ed's poplar barn

Log addition built of pine thinnings

Of course, for most of us, a woodlot consist of more than dead and/or inferior trees. In any woodland management plan these trees would be culled out, giving room for healthier growth of quality trees. To produce healthy, straight, vigorous growth in a woodlot, the woods periodically must be thinned, especially in an even aged, dense softwood stand.

Thinning softwoods can provide some of the straightest, high quality building poles possible.

Generally, when one considers building a home or barn, it is not culled, damaged or inferior trees that one wants. The trees are carefully picked and selectively harvested. A large structure can require some very large, straight trees. If the woodlot was carefully managed in the past, one is in a great position to benefit from the foresight and work of the woodsman who thinned, pruned the woodlot and culled from it diseased, crooked and broken trees.

Jim Lattin approached me with plans for a barn. He asked if I could work with him hewing timbers, cutting joinery and raising the frame. He had a small woodlot, maybe 5 acres in size. He invited me over to walk the woodlot to see if the land had sufficient trees for the barn he wanted to build.

Hewing pine timbers for Jim's barn

I was awestruck by what I saw. Beautiful, tall, straight pines towered above the canopy with considerable hemlock growing below.

The story goes the woodlot was harvested years before but the straight young pines were left to grow in case needed at a future date. Jim, having purchased the property, was the beneficiary of another's wisdom, and the trees that had been left to grow had matured. We felled and hewn some of these pines to create the massive timbers needed to build the barn. The barn was attached to the house, built on the spot where the original barn once stood. Perhaps this is the essence of good stewardship and Jim should complete the cycle by fostering some of the young pines for when again needed many years from now.

Jim's barn

Generally, when hewing, I prefer to harvest trees whose sizes are appropriate for the beams I need. If a sawmill, instead of a hewing axe, is used, it is not so much of an issue. The saw can, in addition to sawing out the timber, saw out boards or planks from what would have been waste.

Occasionally, either by accident or circumstance, I have had to contend with a greatly oversized tree for the timber sizes needed.

Hewn timber framed sauna

Hewn 25' timber 17" square

Once I had contracted to hew, cut and raise a frame for a Sauna complete with changing room and a porch. The overall footprint of the building was to be 18' x 24'. Excavation around the grounds required that a very large, very old pine be dropped. Scott, who I was working for, asked if I thought I could use that big pine in the building. I said that if he had to fell it, I would find a way to use it. Growth rings revealed the tree to be well over a hundred years old. The grain was tight with much of it being resin filled heartwood. Around here people refer to it as pumpkin pine and it is highly valued, far more valuable than young second growth pine.

The largest timbers I needed for the sauna were 8"x10"x24' timbers. The bottom 25' of this log would yield a square timber 17"x17". That is what was hewn.

Ripping timber with chainsaw

Once hewn, new lines were drawn on the timber ends, dividing the log into 4 timbers: two 8"x10"s and two 6"x8"s. Chalk lines were snapped and the timber was ripped into four timbers with the chainsaw. Once ripped the chainsawed surface was lightly re-hewn to make it match the other surfaces. The extra inch in each direction allowed for the saw kerf and re-hewing..

This was followed by a little more cutting, ripping and hewing. The end result:

    Two – 8"x10" x25' principal purlins,
    Two -- 6"x8"x12' rafters,
    Six -- 4"x6"x6' braces.

The process from log to timbers took 2 days.

All these timbers were derived from the 17" square timber.

Of course if one is highly motivated, has strong artistic vision, and has access to a great variety of woods, a very competent sawyer, and a timber framer, the possibilities seem endless.

It is easy to have a pre-conceived idea of what a good timber tree should look like. I picture the tree straight and tall with a fairly good crown and a minimum of lower limbs. When Robert hired me to cut the joinery for his Cruck frame, many of my pre-conceived ideas collapsed. Robert, the sawyer and I would comb the woods looking for curved, bent, crooked and forked trees. Large burls were treasured rather than avoided. The result is a very organic appearing structure.

Another pre-conceived idea of mine that collapsed was how to care for trees. The goal is to thin a woodlot, to open up crown space for the selected ones. In giving them more canopy space, the tree can grow faster, increasing the trunk diameter.

Being a timber framer, there is an advantage to keeping the diameter of the trees from growing too fast. I have always allowed the canopy to remain a little tight slowing the growth rate down. This permitted me to keep the trunk diameter to a more useful size. It also meant, that since the growth rate is slowed, the annual rings are tighter, giving a stronger timber. The risk is that if the stand is kept too dense, the crowded stand may be stressed and vulnerable to disease. This is particularly true with a monoculture stand.

In Norway, to my great surprise, I learned the value of deliberately stressing certain species of trees to increase the durability of the logs used in house building.

Norway has a rich culture of log building, going back to the early Middle Ages.

The Norwegians early on learned that if pine was severely stressed a few years before being felled, the sapwood would diminish, while the resin filled heartwood would increase. The pine would be stressed by having vertical strips of bark removed from the tree. The tree was never totally girded for that would have killed it. The goal was to wound the tree and leave it like that for as many as seven years. When harvested, the resin filled heartwood would comprise most of the log. This wood proved to be very durable. It was both insect and fungus resistant.

Open air Museum Lillehammer, Norway

This method does not work for all species. Norwegians also use a lot of spruce in log work. I was told that stressing Spruce accomplished nothing. It is not in the nature of these trees to become resin filled.

An intimacy with wood and a deep almost intuitive understanding of its properties and limitations is necessary in hand crafting.

At one time in Norway, this understanding of wood's properties was required of a carpenter. Each swing of the axe, each hit of the chisel, each pare with a slick or plane imparts this understanding. Each effort at hand-crafting only deepens this intuitive wisdom. The hand-crafter has a relationship with trees and wood that is unique.

Over the years I have had the opportunity to work with many different species of wood. For what it is worth I will pass on my observations:

Apple: Very tight grained, dense hardwood. Strong and beautiful. The heartwood is a rich, dark brown and contrast sharply with the lighter colored sapwood. It is difficult to find long straight pieces. Makes attractive, functional braces.

Ash: Very strong, light colored, tight, straight grained hardwood. Prone to fairly severe checking and shrinking. The straight, tight grain makes this wood an excellent wood for trunnels (wooden pegs used in timber framing.)

Birch, Yellow: Strong, heavy, cream colored wood. Little to no color contrast between the heartwood and sapwood. Grain lines not very pronounced. Sap has wintergreen smell.

Cherry, Black: One of the more beautiful woods to use in timber frames. Rich reddish-brown heartwood.

Cherry

Cedar, Eastern White: Very light. Easy to work. Good resistance to decay. Not very strong as far as carrying capacity is concerned.

Cedar, Eastern Red: Beautiful rich purple heartwood in contrast to the light colored sapwood. Rot resistant and aromatic. Very knurly wood with a lot of knots. Difficult to work.

Hemlock, Eastern: The preferred timber framing wood in Maine. Most barns and houses in my area were framed with Hemlock. Fairly strong and rot resistant.
Prone to Shake – the separation of the wood along the annual rings. It is also a very splintery wood.

Fir, Balsam: A soft, cream-colored wood, very easily worked. Not very rot resistant. Similar in strength and weight to White Pine, but more brittle. Almost indistinguishable in appearance to Spruce. Aromatic but pitchy when green.

Locust, Black: A very dense, hard, extremely rot resistant wood. If one has access to it, this is the wood of choice for sills. A yellowish-green wood with well defined grain lines. Hard on saw blades, but works clean with chisels. (See *Planning*)

Pine, Eastern White: God's gift to timber framers. Pine is stable. Easily worked. Shrinkage is minimal as is checking. It ages beautifully. It is light, fairly rot resistant, and reasonably strong. It is very pitchy when freshly cut, gumming up, particularly, the measuring tools. If cut in summer it easily develops mildew stain. The stain does not affect the timber's strength, just its appearance. I often avoid mildew stain by oiling the timber as soon as it is dry to the touch. Some question this procedure but to date I have had no problems with it. If the felled tree is left sitting around with the bark on it, Pine borers will lay eggs under the bark and the larvae will tunnel holes through the wood as they develop. Once sawn, hewn, or peeled, it will no longer attract borers. **My wood of choice.**

Pine, Yellow: Very similar to white pine. Usually plantation grown in the Southeastern states. Wide growth rings.

Poplar, Quaking Aspen: A pioneer species. Short-lived. White colored wood. Rots very easily. Unique *cat's eye* knots. Very heavy when green, light when dry. Easily worked. Can develop reaction wood when it has grown under stressful conditions. Reaction wood will snap, crack and/or split severely when the stresses are released either by sawing or cutting joinery.

Oak, Red: Heavy when green. Very strong. Works well with tools. Not nearly as rot resistant as is White Oak. Red Oak is a good wood for Trunnel making. Most Commercial trunnels are Red Oak. The crucks in Robert's frame are Red Oak.

Oak, White: Very strong. Very rot resistant. Another good choice for sills. Works well with tools. Very similar to Red Oak as far as working is concerned. Far easier to work green than dry. Clearly defined grain lines like Red Oak, but whiter in color than Red Oak.

Spruce: A strong, lightweight wood with long fibers. Not very rot resistant. Knots can be very hard and difficult to work. With the exception of the knots, the wood is easily worked. Has a tendency, in some instances, to grow in a spiral twist. Any wood that has a left-handed spiral twist should not be used. It is unstable and will continue to twist over time. Spruce makes a strong, light frame. If one wishes to raise the frame by hand rather than using a crane, Spruce may be a good choice. It is commonly used in Norway in log building.

I am sure someone else could add several species to this list as good building woods, but these are the woods with which I am familiar. I have no experience working with woods west of the Rockies, though I know Douglas fir is considered to be one of the finest choices for both timber frames and log structures. Regardless of where one lives, local woods are generally available. Research and experimentation are the methods to finding which local tree species will meet one's own needs.

Often an inferior species with little market value serves very well in timber frames. The Dutch settlers along the Hudson River built very impressive, wide barns in which the principal tie beams were Basswood. Basswood is a fast growing, very light, weak wood. The Dutch, however, were able to create very large, very long timbers with it and some of the barns built with basswood tie beams are 200 years old.

The English, when they first arrived in North America, sought out White Oak with which to build their house frames, because that was the closest wood to English Oak, the wood used for framing in England. With time the settlers realized the straight tall coniferous forests could meet their building needs much easier.

The hand-crafter has little need to import wood across the continent or across the oceans. One simply has to be aware of what naturally and readily grows in a locality. Learn about its characteristics and appropriately incorporate local species into one's shelter.

Hewing with the Broad Axe

    Hand hewing is an integral part of my life. Every structure on my property is a testament to my dedication to hand hewing. When I decided to go professional my mission statement seemed obvious: "Dedicated to the Preservation of the Craft of Hand-Hewing with the Broad Axe and to the continued use of Hand-Hewn Timbers in construction."

    The most frequent question I am asked is do I use an adze. There is a persistent misconception that most hewing was done with the adze. The reality, however, is that through the Middle Ages to the present hewing has generally been done with the axe. According to Heavrin, *The Axe and Man*, once iron axes became available, the axe replaced the adze as a hewing tool.[1] Hewing axe patterns have been diverse through the years. Certain localities developed patterns unique to a region and

---

[1] Heavrin, Charles A. *The Axe and Man*. The Astralgal Press : Mendham, New Jersey. 1998. p.127.

forest. Other localities fused together axe patterns of different cultures, creating patterns that best met the cultural and practical needs of people living in a specific place and time.

The first issue that must be addressed in hewing is the log itself. Here in North America we are blessed with forest. With the exception of the Great Plains, virtually every area of the country has trees that can meet the needs of the traditional builder. In the Northeast we have oak, ash, hemlock, red or white pine, fir, and spruce. The Southeast has perhaps the greatest diversity of trees available. In addition to the yellow pines, which are prolific, the South has cypress, a variety of oaks, tulip poplar, hickories and walnuts, to name a few. The West has the Lodgepole pine, Douglas fir, Sugar pine, Sequoia, and Aspen. As we move into the northern reaches of the continent we see the greatest diversity and concentration of coniferous trees, all of which can be used as building timber.

Personally, I have hewn various pines, hemlock, spruces, cedars, balsam fir, aspen, oaks and ash. Each tree has its own peculiarities and each requires minor adjustments to approach in regards grain and knots. Any tree straight enough and relatively clear of live knots can be hewn and create a serviceable timber. In years past a favorite tree was the American Chestnut. This tree had grain that could be easily worked with hand tools, yet was very strong and rot resistant. The tree was lost to blight as a timber tree, but the ancient stumps continue to throw out sprouts and it is rumored that some of the stump sprouts are beginning to develop a resistance to the blight. Perhaps someday the chestnut will again be a tree within our forests.

If one intends to hew, tree selection is important. If one is felling ones own trees, selection is easy. If one is purchasing logs, caution is in order. I was talking to the timber hewers at historic Williamsburg. They had some beautifully hewn, massive timbers. I commented on the quality of the timbers. The hewer said that it used to be much more difficult before the hewers themselves were involved with tree selection. When logs are purchased sight unseen, the logs may be over or under-sized, have many large, unworkable knots, spiral twists, or have grown under severe stress that releases itself in extreme cracks and breaks as it is hewn.

I was contracted by the South Carolina Parks system to hew timbers for a reproduction of a 17th century indentured servants cabin. The person purchasing the logs had no experience in hewing. He purchased from two different sources white oak and eastern red cedar logs. The white oak logs proved to be straight and relatively clear. These made beautiful timbers. The red cedar on the other hand had convoluted, knurly grain. The logs were prolific with large, almost unworkable knots. I rejected several of the cedar logs. The remainder we struggled through. The rejected logs were replaced with young straight shortleaf pine, which hewn like butter. Straight with few large, live knots is the rule if one wants to hew. Excessive knots not only make a timber difficult to hew, they weaken the structural integrity of the timber.

Size is another factor. An undersized log will yield a timber with considerable wane on the edges. I generally accept some wane as a matter of course in hand hewn timbers, but I like to keep it to a minimum. Some people like naturally curved edges on their timbers. In the right applications they can enhance appearance. In the 17th century, house timbers were often planed after hewing and the corners were planed with a large round bead. I believe this may have been done to dress out blowouts and wane from the corners. If you do not want a lot of wane, correctly size your timbers.

The other extreme is the oversized log. The oversized log will give one a nice square timber. An excessive amount of labor, however, will be needed to create that timber, and a considerable number of board feet will have been wasted in the process. The hewing process requires an immense amount of human energy, and unless you are one with boundless amounts of it, you will probably want to keep the logs to the desired size.

The method for determining the correct size log is as old as Pythagoras. I use the Pythagorean theorem: $a^2 + b^2 = c^2$. For example: an 8" by 10" timber would require a log that was a minimum of 13" at the narrow end of the log. A reasonable log would be 13" to 15" at the top. Such a log, if straight, should give a nice square timber with no wane. Needless to say, if the timber is to be 40' long, as may be the case with a barn tie beam, the butt of the log will be huge.

There are 3 steps in hewing: scoring, joggling (also called juggling), and hewing. In the case of smaller logs the joggling may not be necessary. Scoring is slashing and chopping into the log to the chalkline, or, as is often the case today, cutting vertical cuts with a chainsaw into the log up to the chalkline. Joggling is knocking out the chunks between these cuts with an axe. Hewing is cleaning up this rough axed surface, shaving it clean to the chalkline with a hewing axe.

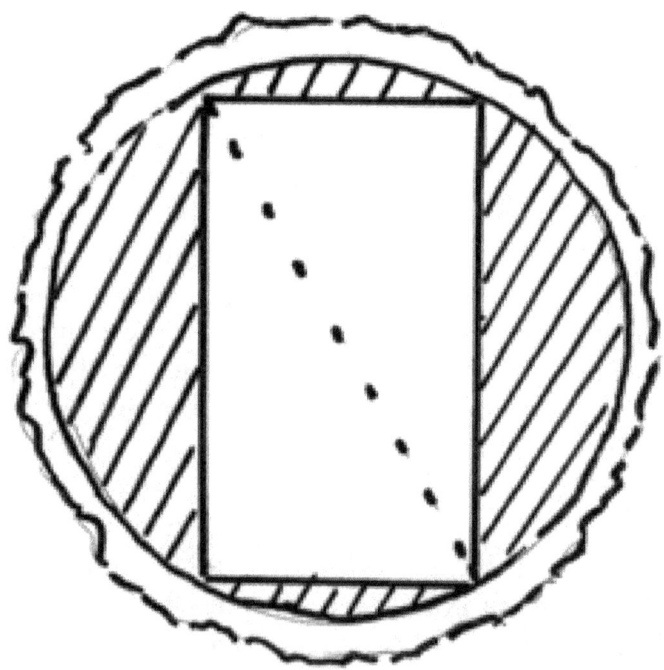

If perfectly square corners are desired, the Pythagorean theorem must be used to calculate the log dimension required.
$a^2+b^2=c^2$
If a 6"X 10" is desired it would require a log slightly over 12" thick at the narrow end.

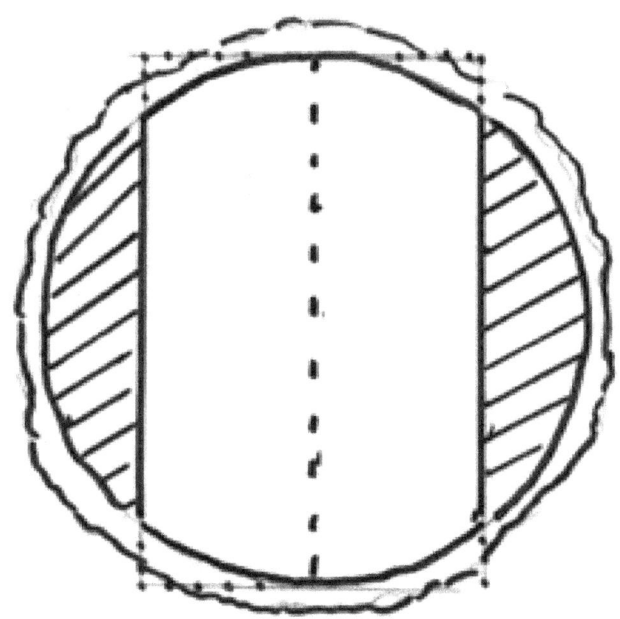

If rounded corners are acceptable, the diameter of the tree at the narrow end need only match the desired width of the finished timber. If a 6"X 10" is desired, the narrow end of the log need only be 10 inches wide.

Large logs are heavy, difficult to move, lift or turn. Small logs are bouncy, unstable and show a tendency to roll while being worked. These create difficulties for the hewer. Available log handling tools and available power for moving and lifting, i.e. people, horses, oxen, machinery, influence directly how the log will be stabilized, scored, joggled and hewn. One with access to ones own woodlot, but with little access to the means to move heavy logs, either mechanical or animal, may find the best place to work the log is where the tree is felled. A finished timber is a great deal lighter than the original log. A third to a half of its mass may have been removed in the hewing process, including the heavy, saturated sapwood. If the finished timber is allowed even just a few weeks in the woods to season, it is surprising how much weight will have evaporated from the timber.

When working logs from my own woodlot, I always hew in the woods where the trees are dropped. Using a small, wheeled, manual log-hauling arch, I can roll finished timbers up to twenty feet long out of the woods by myself.

When I hew for others, what I find can vary. Sometimes I arrive to find the log neatly trestled on high cribbing. More often I find the logs skidded to a general area, sometimes heaped and tangled. Peavies and timber carriers are used to free and maneuver the logs for hewing

Although any attempt at generalizing a process is flawed, I am attempting to explain what I believe was the traditional North American method of hewing in the 18th and 19th century. Photographs taken in the late 1800s show consistency in technique, and the axe marks left on earlier pieces seem to corroborate the same technique. Alternative methods did develop either to accommodate specific needs or individual preference. Some of these alternative methods are covered. I have also included the use of the chainsaw in the scoring process. It is widely used by many hewers today, myself included

The first step in hewing is to stabilize the log. Logs, being cylindrical in shape, are inclined to roll when being struck with axes. To hew a log, this movement must be curbed. The log is generally put on cribbing and dogged into place with log dogs.

One end of the dog is driven into the log; the other is driven into the cribbing. They do a good job of keeping the log from rolling. They are generally placed on the opposite side of the log being worked so that they are not in the way of either snapping the chalkline or working the log with axes. Two generally hold a large log in place. I have found that when working small logs, say a 4" x 6", I sometimes need to drive in additional dogs. The log bounces too much loosening the dogs.

(If working by oneself, it may be difficult to lift a log onto the cribbing. I generally cut a log with a 12" or so diameter to a 3' or 4' length. I than cut this log in half diagonally. This gives me two long wedges going from 0" to 12". I use these as inclined planes and, using a Peavey, simply roll the log onto the cribbing.)

Log Dogs holding log securely.

Assuming we are creating an 8" x 8" timber, the layout procedures are as follows. Once the log is dogged, the ends must be cut flat, particularly if the ends are jagged or uneven from the felling operation. I generally keep the log about 6" longer than the desired length. This gives plenty of play for the joiner to properly square the ends for joining. The dimensions of the timber are then laid out onto the ends of the log.

To do this one needs a pencil, a ruler, and a level. The center point of the log end is found and marked with the pencil. Frequently this point falls directly onto the center pith of the tree, but not always. Through this point are drawn both a plumb line and a level line. Measure 4" from each side of the plumb line, mark the points. Draw plumb lines through these two points. These two lines mark the inside and outside face of the 8" wide timber. Double check that these two lines are in fact 8" apart. Measure 4" up and down from the level line. Mark these two points. Draw level lines through these two marks. These lines mark the top and bottom of the 8" thick timber. Again double check that the lines are 8" apart. Repeat the process on the other end of the log. With the log dogged and restricted from rolling, the outlines drawn upon the log ends define the planes that the hewer must remove.

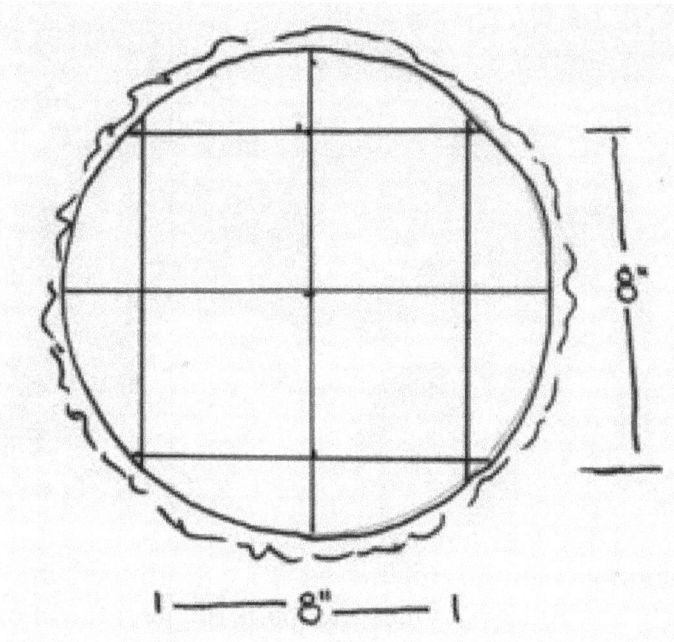

Using Level and ruler the dimensions are laid out onto the log ends complete with horizontal and vertical center lines.

A right-handed hewer generally stands to the right of the log. Logs are generally hewn working from the narrow, tapered end of the log to the butt. The annual growth rings of the log tend to flare out toward the butt of the tree. If one works from the butt of the tree toward the tapered top, there is the possibility for the axe bit to tear out wood along these annual rings, leaving unsightly "blowouts" on the surface of the finished timber. Working from the narrow top toward the broader base, the axe is always shaving across these grain lines, rather than being pulled into them.

A chalkline is used to define plane. Chalkline must be snapped plumb.

One strings the chalkline from the pencil line on one end of the log to its complement on the other end of the log. When snapping the line it is important that one pulls the line up straight, defining the plane to be removed. If the line is pulled out at an angle while snapping, the plane being defined by the chalk may not be the plane one wishes to remove. Always snap the line as plumb as possible. Only one line is snapped at a time. Only one surface is worked at a time.

The next step is to score the surface of the log to be hewn. The word score simply means to mark with cuts or slashes. The scoring process cuts the wood vertically to the chalkline making it easier to remove the wood with the hewing axe.

Scoring with a felling axe.

If the wood were not scored, the axe would bind and the hewer would be incapable of removing the wood cleanly. Traditionally scoring is accomplished with a felling axe. The hewer stands upon the log with feet apart. As the axe is swung, the hewer leans forward driving the axe into the face of the log. The felling axe is used for two reasons. The first is it is made for this type of cutting. The second is the felling axe has a long handle. The minimum size handle of a felling axe is usually 32 inches. This is about as small a handle as one would want in order to be able to reach down deep enough into the log for scoring. Shorter handles simply do not have the reach to be able to cut the lower portion. For scoring in this manner, a 36" handle works best. Often, the best way to get a 36" handle on an axe is to purchase a double bit axe. These often are sold with full length handles. In the 1800s it was not unusual for one to put an exceptionally long handle into the axe used for scoring. Thoreau, in his travels through the Maine woods, describes seeing a scoring axe with a handle a foot longer than normal.[2] If a short-handled axe is used, stand behind the log rather than upon it. This helps one reach deep enough down the face of the log for scoring.

An axe, if it is to have any cutting force, never strikes the surface straight on. A slight slant should be given to the axe cuts. Thirty degrees or so works well. If two inches or less are to be removed, these slanted cuts are all that are needed. They should be spaced every two to three inches along the entire length. Novice hewers frequently score too lightly. For scoring to be effective the entire face of the log to be removed must be cut to the depth of the chalkline. A good scorer is not afraid to reach down and strike hard.

If more than two inches of wood are to be removed, the scorer cuts out wedge shaped openings in the face about one foot apart. To chop out a wedge that penetrates to the chalkline, the opening chopped will be approximately twice as wide as it is deep. This simply is hard work. A sharp axe is a must. Even in a frigid Maine winter, while doing this work, the coat, sweatshirt and flannel shirt are quickly shed to keep from overheating. Once the wedges have all been cut, the resultant surface appears jagged with the uncut wood between the wedges still intact. These jagged protrusions are called jogs or joggles.

In the 1970s Peter Gott, a renowned builder of traditional Appalachian hewn log homes, revolutionized hand hewing. For scoring, he substituted the felling axe with the chainsaw and halved the time it takes to hew a log without in any way changing the finished appearance. His methods differ from mine, but he influenced a change in my own methodology. I have incorporated the chainsaw as the primary scoring tool in my own work.

---

[2] Thoreau, Henry David. *The Maine Woods*. Penguin Books : New York. 1988. p. 30.

Scoring with axe: cutting out wedges approximately a foot apart. Wedge cuts are twice as wide as they are deep. The jogs are still intact.

Jim Lattin scoring with a chainsaw.

If the log is too big to straddle, one can stand on the log while scoring with the chainsaw.

The exceptions to using the chainsaw for scoring are when I am doing a historical re-enactment or a demonstration of traditional hewing. When working with the axe, a wedge must be cut out because that is the only way the axe can penetrate to the chalkline. The chainsaw has no problem penetrating to the chalkline with but a single vertical cut. Generally I straddle the log, hold the chainsaw with the top handle toward the log and the bar pointing straight down. I keep the bar as plumb as possible and cut straight into the log to the chalkline. I am cautious never to let the nose of the saw tip forward or the underside will be cut too deep, leaving chainsaw marks on the finished timber. I place the cuts about six inches apart, working my way up the log. The time and effort saved are immense.

Whether the log was scored with the axe or the chainsaw, the felling axe is also the traditional tool for joggling. The joggles are struck with the axe close to the chalkline. Considerable force is required to split off the joggles. I consider joggling the most dangerous part of hewing. It is because of the inherent risks involved with joggling that I always wear steel-toed shoes when hewing. What makes joggling the riskiest part of hewing is that as the jog is split off, the axe keeps traveling through. If, as the axe travels through, ones foot or leg lies upon its path, one will get cut. Vigilance is the key to safety. One must be aware of where ones feet and legs are when swinging the axe. A continuous, conscious effort must be made to keep oneself out of harms way. Where the feet are placed, how one swings the axe must always be working in unison. Inexperience, working too fast or fatigue increase the risk of injury. Caution, taking ones time, and allowing for rest decrease the risk of injury. It takes but a second to lay open ones foot or leg.

Perhaps the safest way to remove the joggles is to stand on the other side of the log and swing sideways. It is also the least effective and most fatiguing in removing the joggles in a timely and efficient manner, unless the log in rolled back 30 degrees or so to facilitate this sort of axe work.

Other methods of joggling are more effective. Ones personal preference determines which method is right. Some people find it easiest to knock off the joggles while standing upon the log. Others adopt the same stance that they would use in hewing, either straddling the log or standing with the log to the left. If standing on the log, both feet must be behind the chalkline at all times. If standing to the side of the log with the log to ones left: keep right hand in front of left hand on axe handle, keep right foot forward and about twenty inches away from log, keep left foot back and against log.

In joggling the axe is raised high and struck hard, and it takes repeated strikes to remove the jogs. After joggling, a half-inch to an inch of wood is all that should remain to be hewn off. One may find it easier to rescore lightly here and there with the felling axe, especially if more than an inch of wood remains. Frequently, this is not necessary and the hewing process can begin. I find using a double-beveled hewing axe of around 3 lbs to be more effective than the felling axe in removing the joggles. The wider blade seems to be a plus in this operation. Experiment and find what is best for ones self.

Juggling saw scored log, standing to side of log.

Juggling axe scored log.

People frequently speak of the Broad axe as the tool for hewing. Broad axe is simply a generic term that really means nothing more than a broad blade. Hewing axes all have blades broader than a basic felling axe, but some hewing axes have cutting bits not much wider than felling axes, while other hewing axes have bits three to four times as wide as felling axes. Hewing axes can have bits as slight as 4" or 5", or they can have bits greater than a foot or more in length. A great misconception about Broad axes is that they are always single beveled, meaning the bit is beveled on a single side like a chisel and flat upon the other side, rather than beveled on both sides like a felling axe.

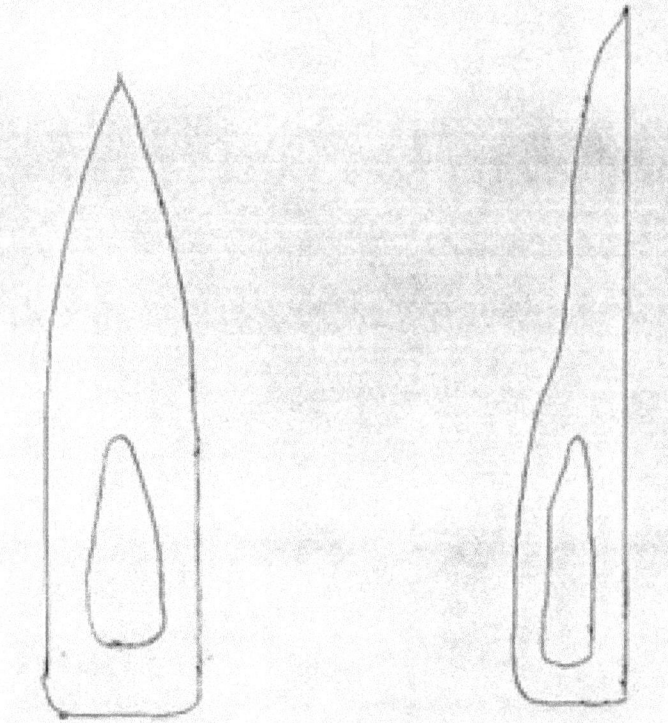

Left: Double bevel – Right: Single bevel

Through the Middle ages right on to the present it is possible to find hewing axes that are beveled on both sides. Both types of hewing axes, single beveled and double beveled, have coexisted for well over one thousand years.

Handle lengths on hewing axes have varied both in time and locale. In parts of Europe, hewing axe handles reached the length of four feet. In North America that length may be as little as 15". Twenty to twenty-two inch handles appear to have been common for much of 19th century America. I keep virtually all my hewing axe handles this size.

Proper stance for hewing: Right foot forward and away from log and right hand forward on axe handle.

The traditional stance in hewing is to stand with the log to one's left. The axe is held with right hand forward, eight inches or less from the axe head. The thumb can either wrap around the handle or be placed on top of the handle parallel with the handle. The right foot is placed forward, the left foot back.

Keeping knees loose helps reduces back fatigue. (For a left handed hewing axe, everything is reversed.) Gentle strokes of the axe are used to sever the line. Once the wood begins to separate, more forceful swings can be used. The broader the blade the more important it is to strike the wood close to near level and plumb. The narrower the blade, the more possible it is to strike at a steeper angle. When hewing, with each swing, allow the flat of the axe to ride along the hewn surface until it hits what is still to be hewn. The axe swing is circular. The axe is raised, swings forward, rides along the flat hewn surface, hits un-hewn wood, and the stroke is finished with a slight slicing motion, as it is raised from the cut. This is the secret of a smooth surface. The slicing motion, opens the cut for the next stroke while keeping the axe from binding in the cut. These motions are repeated as one moves down the log. Gentle swings open the cut along the line. Heavier, slicing motions smooth the face of the log.

Start plumb and to the line.

It is imperative that one starts plumb and that one checks periodically that one is keeping plumb. This can be done with a level. The easiest way to check, however, is to stand tall, align oneself with the chalkline and eye if wood is standing out beyond the chalkline. The tendency is to let the bottom edge jut out. With care and watchfulness, this can be eliminated. Knots are always a problem. The broader the axe, the more difficult it can be to remove them cleanly. The method I have learned from experience is to strike a knot directly in front of it, directly behind it and then, with the final swing, directly on top of it. The bigger the knot, the more force will be required, and the more these three strokes will have to be repeated.

Keeping the work close to waist height reduces back fatigue.

Once the first side is done, check along the length for any high spots and rehew. Transfer the dogs to the other side, one at a time so that the log doesn't roll while transferring the dogs. Simply repeat for the other side. On one side of the log, one moves backward down the log as one hews. On the other, one moves forward up the log as one hews. In this way one is always working from the narrow top of the tree toward the butt.

When scoring and hewing the last two sides, stability is no longer a problem, because the newly hewn surface is sitting flat against the cribbing. Make sure the cribbing sits level so that the line to be hewn is plumb. Scoring the last two sides is easier because the log has been reduced to a common thickness. It is also easier to stand on a flat surface while scoring and one does not need to bend forward quite so much. I have found when scoring the last two surfaces, the chunks frequently fly off in the scoring process and far less needs to be removed in the juggling process.

Bill straddles log while hewing.

A considerable number of people find it easier to straddle the log when hewing. The larger the log, the more difficult this becomes, but the popularity of this stance leads me to believe that it must have been fairly common practice years ago.

Though I describe a standard method of standing and holding the axe while hewing, a standard simply means that this was the *common* method of hewing as left by historical evidence. There have, however, always been variations that defy this standard. There are people that score and hew standing on the log. A long handled axe is almost a must. The angle at which the axe strikes the wood is more likely to be at between 30 and 45 degrees, rather than plumb.

Most of the photos from the nineteenth century in commercial hewing operations show the hewer standing to one side of the log. For the Canadian hewn timber trade, a hewer would stand on each side of the log, one working with a right beveled axe, the other working with a left beveled axe. Together they would hew two sides of the log.

Square Timber Hewing. Archives of Ontario.
(Photo attributed to W. D. Watt, 1912-1913,
Acc. 11778-4/ s16944.)

In *A Book of Country Things*, Walter Needham describes how his grandfather stabilized a log for scoring and hewing. His grandfather would fell the tree and leave the tree crown attached. Charles McRaven describes using the same practice. I also adopted this practice for many years. I would place a small log down where I wished the tree to fall. I would than fell the tree, aiming for the log. If all went as planned, the trunk of the tree would land upon the small log, keeping the trunk out of the mud and dirt. I would remove only the limbs that were in the way of my hewing, keeping the others along with the crown attached. Keeping the upper limbs and crown attached provides stability to the log. It cannot roll and is more stable than if it were cribbed and dogged. I would hew the two sides of the log, cut it free from the crown and then hew the last two sides. Since there is no plumb line to start with, extra care must be taken that one is starting plumb. A level check is a good idea here. I am sure this practice in remote areas predates the hewing of railroad ties. It appears, however, the practice may have been common while hewing ties. Needham states his grandfather used to hew ties.[3]

No Dogs required, log is stabilized by keeping crown attached.

---

[3] Mussey, Barrows. *A Book of Country Things told by Walter Needham*. The Stephen Green Press : Lexington, Mass. 1965. pp. 95-97. McRaven, pp. 74-75. Sloane, Eric. *A Museum of Early American Tools*. Ballantine Books : New York. 1974. pp. 18-19.

There are hewers that stand on the other side of the log, reaching over the log as they hew. Although this is the safest way to hew, I never considered it too practical until I saw a Japanese temple builder hewing. He stood with the log to his right, hewing the opposite side of the log. These Japanese builders have a long apprenticeship. Each worker learns a specific method of working and is expected to work meticulously in this fashion. The body is held in a specific stance, the axe is held in a specific way. How the body is pivoted, how the shoulders are held are all learned. The axe had a full length handle and struck the wood at a sharp almost vertical angle. It is always enlightening to see how a different culture approaches axe work. The tool, the method, the stance are in striking contrast to the general method adopted in North America.

My early hewing stance.

I am a self-taught hewer and when I first learned to hew I developed a hewing method that I thought was fairly unique. I used a long handled, double beveled hewing axe. It had a five inch cutting bit, but the axe head was long and fairly heavy for its size. I would stand with the left foot forward and on the log with the right foot back. I would place the left hand forward on the handle and pivot my trunk while I hewed. Leaning slightly forward as I hewed, I would get an excellent view of the surface I was hewing. I struck with the axe at a rather sharp angle and removed lots of wood quickly. If in a hurry, the scallops left would be fairly pronounced and the surface was anything but smooth. If, however, I strove for a good surface, the hewing was smooth and little different from the hewing I do today. Backaches were common. Research and practice led me to change to the more traditional method.

Man hewing on 15th century print entitled Noah's Ark.

A few years after I had changed my hewing method, I chanced across 15th century prints and paintings of timber construction. In both the hewer was hewing exactly as I had been hewing. In one entitled Noah's Ark, the hewer is standing as I described above (left foot forward and on the timber, right foot back). He is using an axe with the head longer than the cutting bit is wide, and his left hand is in front of the right on the handle. Hewing is this manner virtually requires the axe to be double beveled and the handle to be straight.

I have come to the conclusion that whatever one does with the hewing axe, if it works, it probably has been done before. Axe patterns, as do axe helves, change over time and over regions. The changes reflect not only changes in tool technology but also changes in woodworking methodology. What I describe as the traditional method of hewing is at best the method brought from England to the colonies and used extensively by North Americans throughout the 18th and 19th centuries.

In the so-called "Viking Age" axes were used for even more woodworking functions than the 18th century. By the 18th century, if planks were needed, a large timber would be hewn, placed on trestles and pit sawn into planks. In the 11th century, the hewn timber would be split with wooden wedges. The rough split plank would then be placed into a tree crotch, wedged tight into place and hewn smooth. The rugged sea-faring Viking ships were made in this fashion. This is depicted on the Bayeaux Tapestry as William the Conqueror prepared to invade Britain.

I know of no one today hewing planks in this method. I have, however, seen 15th century prints in which the hewer is hewing a log sitting on trestles instead of cribbing. And this practice appears to be gaining favor with many hewers. Trestles put the log up at waist height and make it easier for the hewer to work without back fatigue. The problem with trestles has always been how to get the log up there. Logs are heavy and some form of mechanical energy greater than what a person is capable of providing is necessary. Most of the prints I have seen show relatively small timbers on trestles. In this modern industrial age we have machines (backhoes, skidsteers, tractors) capable of lifting heavy logs and easily placing them on trestles. Peter Gott, who strongly influenced the use of the chainsaw as a tool for scoring in modern hewing, also created a means for getting timbers onto trestles for hewing without relying on machinery. His method was simple: score and joggle each side before placing the timber on trestles for hewing. Using Peter Gott's method, the timber is laid out while sitting on the ground. Lines are snapped. Horizontal cuts with the saw are made through the top surface. The log is pivoted 30 degrees and joggles are knocked off by the hewer who stands on the opposite side of the log. This is repeated for each side to be hewn. The remaining timber is much lighter, having had most of the waste wood removed. Using a couple of timber carriers, the timber is placed on trestles. Since no more scoring is required, the hewer goes directly to hewing each surface of the timber. [4]

Regardless of the method, the important things to remember with a hewing axe are secure the timber, hew plumb and finish each stroke of the axe with a slicing motion. These three things help insure a square, smooth timber. In learning to hew, the hewer goes through stages of development. Some pass through these stages quickly, having an intuitive grasp of tools, their body and wood grain. Others, like me, have to painfully and slowly learn these stages. The first stage is how to hold the axe and position one's body. If one grasps the handle too tightly, the fingers will cramp into the clutched position even after the axe has been put away. One eventually adopts a posture or stance when hewing that is not so taxing on the body. When first learning to score, joggle or hew, the body is very tense. Back, forearm, shoulder and hip aches are all common. One must learn to relax the body while hewing, tensing only the muscles needed to perform the tasks. If one continues to experience pain, perhaps the stance should be modified. Put some spring into the legs. Find ways to ease the strain on the back. Take frequent breaks. Walk around. Give your body a rest. Developing ones skills with the tools helps go a long way toward relieving strain on the body. Carefully placed strokes relieve strain. Carefully placed strokes minimize risks. Knowing where ones body is in relation to the axe and knowing where the axe will strike is also the essence of axe safety.

The second stage of learning is all about the axe. Every axe has a different weight, length, bevel, offset to the handle, and handle length. With each axe one must learn to adjust to all of these factors. If one hews with more than one axe the learning must occur for each axe. Even after years of hewing, I have to pay attention if I switch axes, adjusting my method of hewing.

---

[4] Langsner, Drew. *A Logbuilder's Handbook*. Rodale Press : Emmaus, PA. 1982. pp. 96-120.

The variations of the width of the cutting bit, the bevel and the offset are the factors that are the most challenging. These factors must be adjusted for or the axe will simply deflect off the wood or produce ugly gouges and cuts into the finished surface. When teaching people to hew, I usually tell them to pay attention to the bevel of the axe. Watch how the bit is striking the wood. Make adjustments to the stroke as one goes. When the bevel of the axe is understood, it makes no difference whether we are talking single or double beveled axes, the axe leaves a clean hewn surface. It becomes intuitive; one adjusts without thinking. The axe slides effortlessly along the hewn edge into the cut, leaving no slash marks.

  The final stage of learning is understanding the wood grain itself and the approach to the grain one must take to smoothly hew the timber. This is a progressive knowledge acquired over time. It is perhaps the most intuitive of all. Every tree species has different grain. The wood grain can be long and coarse or short and smooth. It can be very straight or it may be twisted and interwoven. What works easily for one species may not work for another. Clear Eastern White Pine logs are perhaps the most forgiving trees one can work. It remains my favorite. It is the wood I generally use when teaching others to hew. Though I usually start hewing at the narrow top of a log, hewing my way back toward the butt of the log, with pine I have hewn from the butt to the top with little difficulty. With Eastern Hemlock or oaks this could be disastrous. The bottom edge of the timber would be inclined to blow out along its entire length, piece by piece as I worked my way up the log. When I worked white oak logs, I found the smoothest, straightest surfaces resulted when I started at the top of the log and worked moving forward toward the butt. To work only in this fashion, the log had to be rolled and re-dogged after every side rather than after completing the two parallel sides. This, for me, kept the lower corner from blowing out. Adjustments are the essence of good hewing. As we experiment with different hewing situations, we can reach back into our bag of experience and pull out different strategies for working with different tree species.

Angle of strike for single-beveled axe.

Angle of strike for double-beveled axe.

Butt flares of almost any species can require that one pulls out the whole bag of experiences. At the butt flare, the bole of the log spreads and begins to differentiate into what are to be the principal roots of the tree. The grain flares out, undulates and sometimes interweaves. Working down the tree, it is important to sever the leading edge of the grain before simply striking the butt. If not, as the butt is struck, it will split out chunks of wood from the finished surface of the timber.

Knots are another issue. Some species, like white pine and aspen have soft knots. It requires additional force to hew these knots free, but basically one does not have to do any major adjustment to the hewing. The knots will hew out clean. Spruce, on the other hand, has very hard knots. Spruce is a very deceptive wood. It is a soft wood and easy to work until one encounters the knots. Spruce knots can take the edge right off a tool. The ideal situation with knots is to score the hell right out of them, removing them practically to the chalkline. I tend to hew the wood off in front of the knot and in back of the knot isolating the knot. Then I strike, repeatedly, down onto the knor. With the wood removed both in front and behind the knot, the risk of blowing out wood surrounding the knot is greatly diminished. Eastern Red Cedar is literally peppered with hard, difficult knots. On extremely knotty wood, it sometimes is best to switch from a wide bladed hewing axe to a narrow bladed hewing axe.

Some species prefer to grow with a slight spiral twist. With a spiral twist, one can hew the top edge moving backward, the middle is no problem in either direction, but the bottom edge must be hewn moving forward. Otherwise the bottom edge will surely blow out. It should be said that severe spiral twist in a log is a reason for rejecting the log. As the hewn timber seasons it will continue to twist.

Whatever I have to say about approaches to grain and knots can only scratch the surface. If a log is bowed, knotty and twisted, you will have to intuitively compile several strategies when hewing the log. One learns grain and how to approach it by hewing. Understanding wood grain may have been knowledge passed on from one generation to another years ago. Today most of that knowledge has to be learned by trial and error.

Essentially it comes down to: know how to work safely, know ones tools and know ones woods.

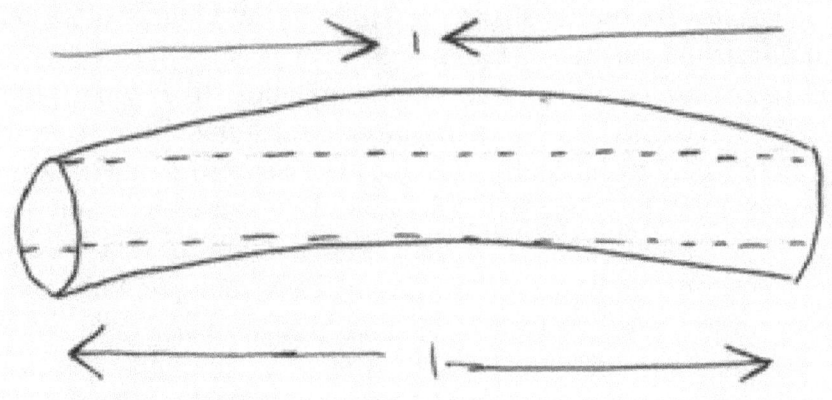

Arrows show direction to hew with bowed log.

Hewn White Oak timbers.

Axes from left to Right: Swedish carpenter's axe, Swedish pattern hewing axe, Pennsylvania pattern hewing axe, and Alpine pattern hewing axe.

Copy of 18th century hewing axe, forged and used at Williamsburg.

Copy of a 12th century Viking hewing axe, forged by Lars Enander, Gransfors Bruks.

Hewing axe patterns vary greatly. Few companies today manufacture hewing axes. The most reputable company today is Gransfors Bruks. Their double beveled hewing axe is a true gem of a tool. It is, however, very expensive. If one wants a large single-beveled hewing axe, the best sources remain antique tool stores and Ebay. I believe the finest hewing axes in the world were manufactured in North America during the latter half of the 19th century.

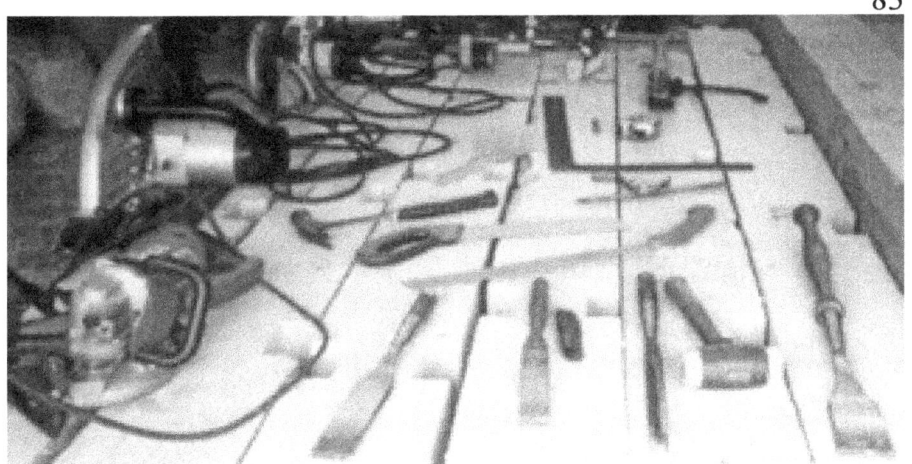

Tools of the timber framer: Left – circular saws. Front (left to right) – 2" heavy duty socket chisel, 1 ½" chisel, corner chisel, 3 lb. split hide mallet and slick. Behind chisels – two Japanese style pull saws. Back – combination square, framing square, tape measure and small cant dog.

Tools

If one wanted, one could easily spend thousands of dollars on timber framing tools. Whether or not this makes sense for one depends upon what one is doing. The needs of a person running a timber-framing factory are considerably different than the needs of an itinerant timber framer. And the itinerant framer's needs are different than the needs of the occasional framer. If one only intends to build but a single frame, the tool needs can be kept quite minimal. This may require using simpler joinery or utilizing more labor intense methods to create the joints, but this is more sensible than purchasing a very expensive tool that will be used for but a few hours in the course of a few weeks, never to be used again.

In 1980, Bill Behrens approached me, asking if I would be willing to work with him on building a hand-hewn barn frame. Our collective tools consisted of the following: Chainsaw, 2" framing chisel, bit brace with 1" auger bit, hand saw, hand made mallet, hewing axe, level, square, sliding bevel, measuring tape, chalkline and a hammer. A circular saw was available, but with the exception of cutting the floor planks, I do not recall using it.

Frame cut solely with chainsaw, chisel, bit-brace and mallet. Simplicity of joints made it possible to cut with not more than chainsaw and chisel.

The finished barn was a 1 ½ story 25' by 36' barn. The timbers were hewn square with a single hewing axe. Most of the joinery cuts and all of the mortises were cut with the chainsaw. For mortises the tip of the chain bar was plunged into the wood. The mallet and chisel were used as clean up tools.

Where a lap joint instead of a mortise could be used, it was. The lap joints were much easier to cut and clean out when the principal tools were the chainsaw and chisel. Neither of us had ever built a timber frame. Lap joints simplified the layout.

The bents were assembled on the deck. Tie beams were laid onto the posts and the half-dovetail was scribed onto the post. Once fitted and re-checked for square they were pegged. The braces were then laid on top and scribed onto tie beams and posts. The brace pockets were cut and the braces dropped into place and pegged.

Keeping the joinery simple enabled us with few tools to cut and raise the frame. A quality, complex frame does not necessarily require a lot of tools. Once when I was at an antique tool store a man came in with a tool chest. He said the chest belonged to his great-grandfather who had been both a ship builder and a timber framer. He was looking to sell the entire chest.

We opened the tool chest and there the tools were. Everything was neatly placed in the chest. A hand operated mortising machine was folded on top. Beneath there was a wooden bit brace with brass inlay and bits, several wooden planes, a small ax, two hand saws (one for ripping, one for crosscut), two framing chisels, a square, a scratch awl, dividers, a folding rule, and a slick.

The carpenter obviously had cared for his tools. There was no rust on any of them. The chest had been stored for over a century in some dry loft, perhaps never opened. Many jobs may have required additional tools but this tool chest was a clear record of the tools this timber framer deemed necessary for his trade.

In actuality, little has changed today. We need tools for boring, measuring, cutting and shaping.

The premier mortising tool today is the chain mortiser. It is, however, an expensive tool. The model shown below, made by Makita, is perhaps the least expensive and it sells for around $1500. Some mortising machines sell for twice that much. The Makita machine is fairly easy to use. It utilizes a double cutting chain, similar to that used on chainsaws. Chain comes in ¾" or 1" widths. The mortiser clamps to the timber. It can be easily adjusted along the width of the timber, and it has a depth adjustment for controlling depth of mortise. Once set it operates like a drill press. After the initial plunge cut, the mortiser pivots, extending the plunge cut. The result is a ¾" by 4" or 1"by 4" mortise up to 6" deep, depending upon the chain being used. With a simple adjustment, the mortiser can be slid along its frame to widen the mortise to 1 ½" or greater.

With this machine it is simple to do mortises any width, any length. For thru-mortises, one simply turns the timber over and plunge mortises the opposing side. It is, however, not a finished mortise. One must still *clean it up* with framing chisels and/or a slick. This mortiser is particularly handy for sawn or planned timbers. The irregular, rippled surface of hewn timbers can pivot the clamping features of the mortiser off slightly, skewing the mortise. With live-edge timbers or round logs, I question the value of the chain mortiser altogether.

Makita chain mortiser   *½" drill with self-feed bit*

Out in the field, the tools I am most likely to use for boring out mortises are a ½" drill and a self-feed bit.

Self-feed bits with their screw tips and multiple cutters will do straight smooth holes. One can even overlap the holes, something a basic auger bit will not let one do. Makita, DeWalt, Irwin and Milwaukee manufacture sell Self-feed bits. I keep on hand 2" and 1 ½" bits, since timber frame mortises in general are either 2" or 1 ½". The Self-feed bits without extensions will bore holes up to 4 inches deep.

Economically there are advantages to this approach. Self-feed bits can be purchased for under $50. They are easily re-sharpened with a triangular file. And regardless of whether one uses a mortising machine or a self-feed bit and drill one will still need to purchase a ½" drill for general trunnel boring.

The bits used for peg boring are called ship auger bits. I keep on hand a 1", a 1 ½" and a 7/8" bits. The trick with auger bits is not to force them. Let the bit go in easily. Sometimes it is even necessary to hold back on the drill. This allows the auger to discharge the wood it is cutting. Pushing down on the bit may enable the bit to cut faster than it can discharge the wood waste, binding the bit and stressing the drill. I generally use 1" pegs but sometimes in the course of making pegs I end up with several undersized pegs. When that is the case, I switch to the 7/8" bit.

The 1 ½" bit is for those occasions where an oversized peg is needed, such as securing an un-tenoned rafter to a tie beam.

Ship auger bits with octagonal or hexagonal shafts will fit a manual bit brace if that is one's preference.

Traditional bits for bit braces can still be purchased from mail order suppliers. Antique tool stores usually have these in almost unlimited quantities.

When it comes to cutting we have a great advantage over the traditional timber framer's handsaw.

Though underrated, perhaps the most useful saw today is the chainsaw. The chainsaw with a properly sharpened chain is an incredibly accurate tool, especially in the hands of an experienced and skilled operator. Timber framers that I know tend to frown on the use of the chainsaw, but in the trade of log building the tool is considered indispensable.

Circular saws also come in very handy. I use an 8 ¼" saw, but a standard 7 ¼" saw will do just fine.

16" Makita circular saw has a cutting depth of 6 ½".

I also use a 16" circular saw. This is a powerful and heavy saw. It needs to be used on at least a 20 Amp breaker. It has a maximum cutting depth of 6 ½". A 6" x 8" can be cut with but a single pass. Anything larger and the piece must be turned over to finish the cut.

Like the mortising machine, this is a specialized and expensive tool. The saw cost more than $600. If one is pursuing timber framing as a trade, the purchase is justified. If one is simply attempting to raise one or two frames, a standard 7 ¼" circular saw and a good handsaw will meet most of one's needs. And if one is good with a chainsaw, the 16" circular saw is not necessary.

A 7 ¼" saw will cut to a depth of 2 ½". An 8 ¼" saw cuts to a depth of 3". A 10" saw cuts to a depth of almost 4".

Regardless of the circular saw one has, one is definitely going to want a handsaw. Almost all of the timber framers I know own one or two Japanese style pull cut saws. I keep two: a fine tooth crosscut and a course toothed one for heavy cutting or rip work.

The quality of these saws can vary considerably. One can pay quite a lot for a saw made by a Japanese craftsman, or one can get a generic saw for around $30. These saws can be re-sharpened with special feather files designed for the saws. If one purchases the generic saws, however, the blades are designed to be replaced.

The advantages of the Japanese style saws to traditional Western saws is, because they cut on the pull stroke instead of the push stroke, the blade is thinner, requiring less effort to move it through the wood. The saw teeth also require less set. The thinner blade with less set creates a much narrower kerf. The narrower the kerf, the less one has to work. If a circular saw cannot cut deep enough, it is no problem to fit this saw into the kerf and finish the cut.

Perhaps the tool that most differentiates the timber framer from the conventional carpenter is the framing chisel. It is an 18" long chisel with a socket, a wood handle and a metal ferrule to protect the wooden handle. These are not cabinet-makers' chisels. The standard sizes are 1", 1 ½" and 2". Since most mortises are either 1 ½" or 2", these are the more common sizes to own. If one wanted to purchase only one chisel, I would recommend the 2". I find one can easily design a frame using only 2" wide mortises. The framing chisel is the workhorse of the timber framer. My 2" chisel is indispensable. I may spend 4 hours of an 8-hour day using it. It is repeatedly struck forcefully with a mallet. It is used to cut, to pare, to slice, to scrape and to pry. Virtually any aspect of any joinery created will have to be worked with the chisel.

2" socket framing chisel being struck with 3 lb. split hide mallet.

Nearly as valuable as the 2" chisel is the 1" corner chisel. For getting crisp ninety-degree corners in a mortise, this is the tool of choice. Like the framing chisel, it is repeatedly struck.

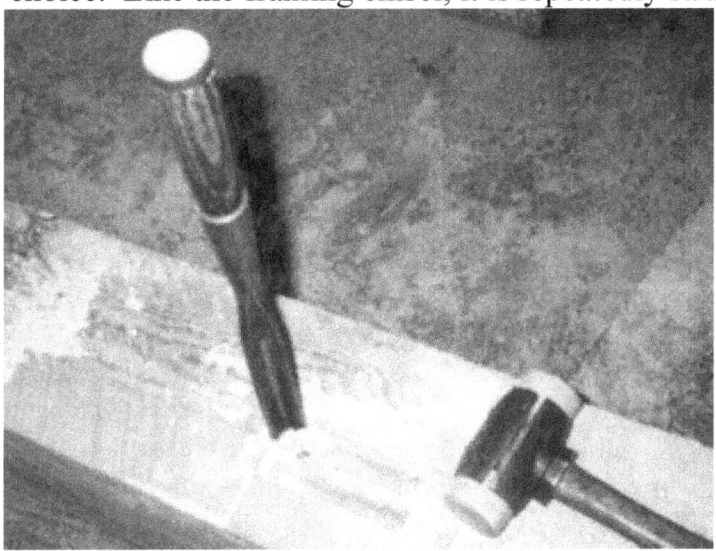

Corner chisel

If the framing chisels were struck with a metal hammer or a small steel sledge, in time the wooden handle would shatter.

Chisels are struck with mallets that can deliver the blow without damaging the chisel.

The simplest tool for striking chisels is a wooden mallet. This can be quickly whittled from a piece of firewood. One can also use a rubber, dead blow hammer.

I find a 3 lb. mallet to be just right. My preferred mallet is a split rawhide mallet. The weight of the mallet is in the cast iron housing that holds the rawhide. The cast iron creates the power of the blow but the rawhide delivers it. This mallet is far superior to either a wooden mallet or a dead blow mallet. It also is the most durable. When the rawhide inserts wear out, they can be replaced.

Of course, chisels are not always struck. Often we use them to pare or shave a surface. For paring down the sides of a tenon or a large scarf joint one may want to use a slick. The slick is like a very large chisel. Usually the blade is 2 ½" or wider. The handle is longer than a framing chisel and has no metal ferrule, because it is designed to be pushed rather than struck. The long handle is usually offset from the plane of the blade so that it does not interfere with shaving or paring.

Usually the chisel and/or slick will create a smooth surface within the tolerances necessary. In certain circumstances, however, I will use a plane, electric or hand. The electric plane is a good tool to keep handy in either log or timber work. This plane can be used to bring a scarf joint to the line or it can be used to shave all saw marks from rough sawn timbers. The electric plane is a hardworking tool. I use them hard and I get years of use out of one. The blades are easily replaced and generally readily available at local hardware stores.

Where measurement and layout are concerned, the most important tool besides the basic tape measure is the Framing Square.

In timber framing there are two common methods of layout that are used most frequently. The first and oldest method is called scribe in which timbers are aligned atop one another in the position they will be in the frame and the points of intersection are scribed onto each other. When properly done, even irregular surfaced timbers can be seamlessly fitted.

The other method is called Square Rule, in which all measurements are taken from a reference face, usually the outside face of the timber. In this method all joinery is laid out using the framing square. The advantage is the entire frame can be laid out and cut without pre-fitting.

A complement to the framing square is the smaller combination square. It is a good tool for setting and checking depth, particularly inside a mortise. The combination square, in addition to checking mortise depth, is a good tool for checking the plumb of mortise walls.

Once we move away from ninety degrees the adjustable, sliding bevel is an excellent tool for both layout and checking.

The other three tools for layout that I find necessary are a set of dividers, a level and a plumb bob. When scribing one timber to another it is absolutely important that the timbers being scribed are level to one another and the scribe lines being transferred are plumb.

The one other measurement tool that rounds out my timber framing tools is a carpenter's calculator. It cost less than $100 and the time it saves in calculating rafter lengths alone quickly pays for the calculator.

Log builders require a few different tools. Not too long ago I went to Norway to attend a two-week course in Norwegian scribe-fit log building at Norsk Lafteskole. I was actually surprised by how few tools were needed for building a tight log home.

Logs were left to season under a tarp for a year. The logs were then gone over with an electric plane, removing mildew stain, knots and any residual material from the cambium layer. The Norwegians will plane a rather large log into an oval shape so that it will neither jut out excessively from the outside wall nor jut too far into the living space on the interior wall. Once planed, the primary cutting tool is the chainsaw. The chainsaw is used to cut, scrape, plunge. The handsaw is seldom even picked up. The other *finishing* tool is the Lofting axe.

Norwegian locking joint

Norwegian lofting axe. With this tool the locking joint is hewn.

    The principal layout tools are a pocket level and a folding rule. The folding rule acted not only as a measurement tool, it was used as both a spacer and a set of dividers.

To create the channel on the underside of the log so that it seats and creates a seal with the log below, the tools used were a hand forged scribe, the chainsaw and last, the channel knife.

Principal layout tools in Norwegian scribe fit log building were the folding rule and the pocket level. Not only were these used to establish plumb and vertical measurements, but were also used as spacers.

Each log builder was provided with a gouge. In round log building, the gouge was far more useful than the chisel.

Other tools found useful for log building were timber calipers, a tape measure and a chalkline. On a few occasions log dogs came in handy for steadying a log being worked upon.

What becomes obvious watching a competent log builder is that skill with a chainsaw decreases the needs for other tools. The tools I have listed here are either the tools that I use most or tools that I have witnessed others using. Each job will slightly vary the list of tools, and tools not mentioned here may be required.

Norwegian Lofting tools: Channel knife, scribe, gouge.

Scribed channel or cope

Fitted logs

Though I have not spoken about straps and come-a-longs, I have yet to see a frame being raised that has not required straps and come-a-longs.

If one is purchasing tools, particularly chisels, purchase the best. Inferior tools are frequently over-hardened making them both brittle and hard to sharpen. A good tool forger knows how to laminate, and temper steel so that it takes and keeps a good edge. The current buzzword in chisels is Barr. Barr Quarton is perhaps the premier manufacturer of timber framing chisels today. His ability to work steel has gained him international recognition.

When available and affordable, I believe it is always better to purchase hand-forged edge tools over drop-forged tools. And, as we move deeper into the world of hand-crafted shelter, it only makes sense that we use hand-crafted edge tools.

Mortise and Tenon

Mortise and Tenon joinery form the basis of timber frame construction. It is what holds the structures skeletal system together. There are post and beam structures that have no mortise and tenon joinery, they are held together with metal fasteners and bolts, but, generally among timber framers, structures such as these are not called timber frames. In their definition, to be a timber frame it must be joined with mortise and tenon joinery.

For one just starting out, timber framing can seem daunting. One may have limited skill, tools and money. The solution, if one is building a first frame, may be to just keep it simple. One can avoid complex tying joints, use simple drop in joists, and use birdmouth rafters. Birdsmouth rafters can be designed such that no notching need be done to the top surface of the plate to receive them. All of these options are described in the following chapters.

In the 80s I had the opportunities to help my brother Den build his camp, my neighbor Bill build his barn, plus opportunities for building upon my own property. Years later I drew from those experiences to create what I believe is the simplest timber frame one can cut. In recent years I have built three such frames. All were sixteen feet wide. The lengths varied.

A basic and simple mortise and tenon frame

In this basic frame, each post has a simple 2" stub tenon for sill placement, a three inch tenon for the plate, two or three brace mortises, and a thru-mortise for the tie beam.

Tie beams have a full width tenon on each end and two brace mortises, each three feet in from tenon. (See *Tying it all together,* and *Bracing.*)

If a loft is desired, joist pockets must be cut, generally 24", 32" or 36" on center. The plates have mortises to set onto post top tenons and mortises to receive braces. If the structure is long enough to receive a scarf joint, the simplest scarf to cut is the stop-splayed scarf. (See *Scarf Joints.*)

When doing timber layout, the first thing one should do is inspect the timber. I check for crown. If it is to be a horizontal timber within the frame, ideally the crown should face up. If it is to be a vertical post, one would want a good flat surface facing out. Such a surface would facilitate sheathing the frame in. The timber should also be checked for shake, rot or any other deformities that may in anyway severely weaken the timber.

If the timber has been found to be acceptable, and crown, etc. have been determined, the reference faces must be marked.

Timbers, generally, have two reference faces. These are the faces of the timber that face the outside of the building, or if it is a horizontal timber, either its top surface, as in a plate, or its bottom surface, as in a sill, will be a reference face. These surfaces are marked as shown below.

Corner post with reference faces clearly marked

All joinery is laid out from reference faces. When laying out a tenon or mortise we are always measuring from the reference face.

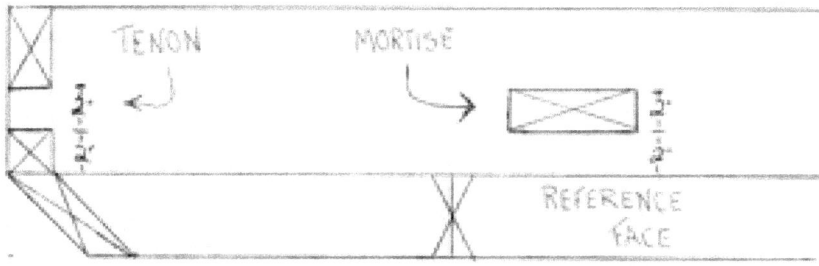

Tenons and mortises are laid out from reference faces. 2" tenons and 2" mortises are generally laid out 2" from reference face unless circumstances make this impractical.

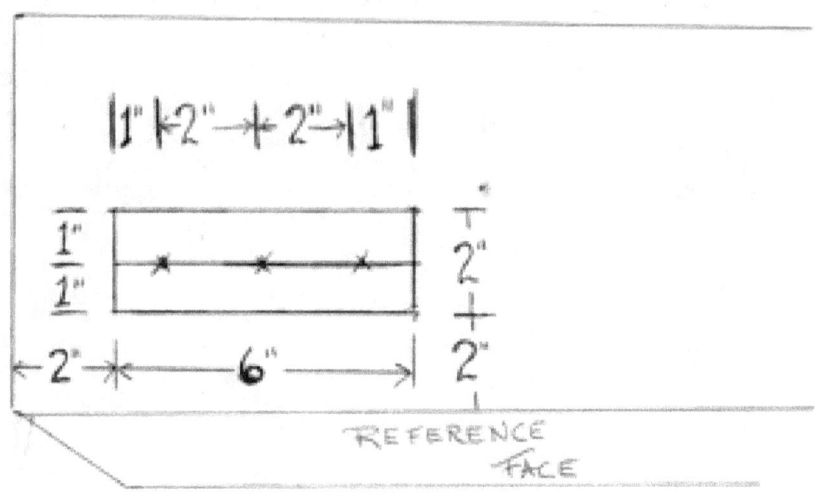

Laying out and cutting a mortise:
Two inches of wood are kept uncut at the edge of a timber. Since the mortise is 2" wide, the mortise likewise is positioned 2" in from the reference face. A centerline has been drawn down the center of the mortise. I bore with an electric drill and a self-feed bit. The centerline is to help center the self-feed bit. The Xs along this centerline are to pinpoint the screw tip of the self-feed bit for boring.

Prior to boring one should strike along the perimeter of the mortise with a framing chisel and mallet. This helps keep wood from tearing out beyond the mortise. When boring, the drill should be held straight and plumb. To check drilling depth, use a combination square set at the desired depth. When the flat of the square can sit upon the face of the timber, the hole is deep enough. Once bored, the mortise is cleaned up with a chisel and checked with the combination square to be certain the walls of the mortise are plumb and to the correct depth. **If the mortise is to receive a one inch trunnel, the minimum depth the mortise can be is 3".**

A finished mortise: 2" wide, 8" long, 3" deep

Cutting a Tenon:

The tenon is always laid out and cut 1/8" shorter than the depth of the mortise. This is to insure that as the timbers shrink, the end of the tenon does not bottom out, preventing the timber from seating tightly against the other. Because the tenon is not centered but rather two inches from the reference face, when performing the cross cuts, the cut away from the reference face will exceed the cutting depth of a 7 ¼" circular saw and the cut will have to be finished wih a hand saw.

Tenon laid out 2" from reference face (right face.)

Do cross cut and rip cuts with circular or hand saw.

Knock off waste with chisel and mallet. If saw can cut to correct depth, these pieces will simply fall off.

Pare clean with chisel.

Check measurements of tenon and check that shoulders of timber are square and true.

If the timber is a corner post, the 2" toward the outside edge are removed.

Chamfer edges of tenon to ease fitting tenon into mortise and the tenon is finished.

There are 3 major methods of laying out and cutting timber frame joinery. The first is to let the dimensions of the milled timber rule layout. We order an 8"x8" timber from the mill and cut joinery assuming it is an 8"x8" timber, or, if it is not, minor adjustments are made in length of connecting girts to adjust for discrepancies. This appears to have been common practice in the area of Maine in which I live. In actuality, sawn timbers are seldom exactly the dimension they are labeled, nor are they perfectly square.

In a small frame, the slight variances in dimension will probably have slight affect upon the overall frame. On a very large frame, the slight variances, if not adjusted for, can accumulate to be considerable. A ten-bent barn growing by ¼" per bent could find itself sticking off the foundation wall by 2 ½" when the final bent is raised.

To avoid this problem timber framers in the early 19th century developed a method called *Square rule* framing. Because sawing is an imperfect craft, in square rule the timber does not rule, the framing square does. Within every imperfect 8"x8" timber lies a perfectly square 7 ½"x7 ½" timber. If, by using the framing square against the reference face, every mortise and tenon were relieved and fitted as if the timber were 7 ½"x7 ½", the joinery would fit perfectly and the structure would remain constant regardless of how many bents the frame had.

Rafters, post and braces in this frame are recessed into receiving timbers. All housings were measured from the reference face insuring a consistent fit regardless of imperfect dimensions of sawn timber.

Girding beams, braces and post square ruled to ½" less than nominal dimensions of timbers. (Work done by Bud Menard)

Square ruled mortise.

This 8"x10" plate varied along its length from a thickness of 9 ¾" to 10 3/16". Square ruling this timber to 9 ¾" insured that all post, and braces entering this timber did so at a consistent height, keeping the top of the plate level. This housing was relieved to receive a corner post and the 3" mortise bored and cleaned.

Square ruled tenon.

As with the mortise housing, the timber must also be relieved around the tenon. This 8"x8" timber had the same inconsistencies as the above 8"x10". To insure that the timber will fit into the housing created, this timber needed to be reduced to a common dimension. The outside faces were left untouched. Measuring from the outside faces (reference faces) the timber was reduced to 7 ¾". Consistently reducing all timbers to a common dimension wherever timbers intersect enables one to seamlessly fit a frame.

In the chapter entitled *Bracing* a square rule brace mortise is laid out and cut. The brace is relieved into the timber one inch less than the nominal dimension of the timber. Though the example given is a brace mortise, the methodology would apply to any mortise cut using square rule framing.

The one other framing method is by far the oldest. The method is called scribe rule.

In scribe rule, neither the timber dimension nor the framing square rules. Timbers are carefully placed one atop the other and using a level, plumb bob and pencil, the timbers are scribed to fit one another.

Scribing easily goes back to the early Middle Ages. People knowledgeable in the history of scribing say that entire frames have been laid out and cut with neither the framing square, ruler, nor measuring tape. A stake driven in the ground with a string attached was sometimes used to determine rafter angles and lengths.

In scribe rule, each timber fits in a specific location and cannot be transferred anywhere else in the building. Braces are always the easiest example. In either mill rule or square rule framing one cuts all the braces identical to one another. When a brace is needed during the raising, it does not matter which one is grabbed for they are all the same. In scribe rule, each brace has been scribed to fit a specific location and will not properly fit anywhere else in the frame.

Scribe rule may be the best method to employ where one is using irregular or round timbers.

Timber positioned to be scribed

Collar tie after scribe fitting

Sills and Floor Joists Systems

Sills and Sill Girder Beams:
Many timber frame companies dispense with a timber-framed sill and joists system, finding it easier to simply do a conventional sill and floor system using kiln dried, dimensional 2" planks. Even where one has decided to use 2" kiln dried stock for the first floor sills and joists, generally the second floor joists are exposed and a timber frame system is employed. I prefer a timber-framed system. It adds to the basic character of the hand crafted home.

Perhaps the classic sill and floor joists system brought to this country from England is the basic Cape. This system involves, in addition to the sills that form the perimeter of the building, sill girder beams that run across the width of the building, one on each side of the central chimney, and two chimney girts tying the sill girders as well as forming the outline for the chimney base. I have seen house sills as small as 6"x8". The sills were laid with the 8" side flat against the granite base leaving but a 6" rise to receive the joinery for post, sill girders and joists. The sills, being supported along their entire length, do not need to be large timbers, but I, personally, recommend sizing them in proportion to the timbers they are to receive. I would not recommend the sill be less than 8" in depth.

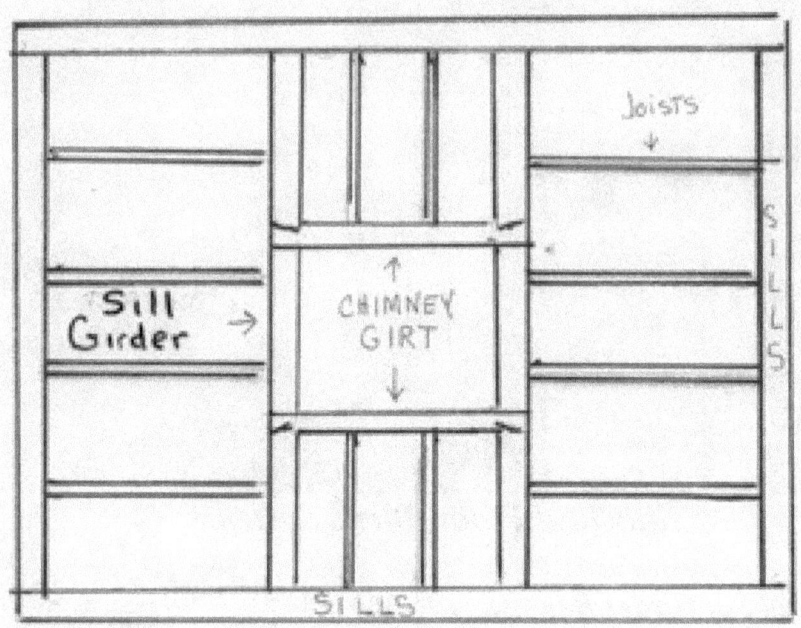

Sketch of a floor system from cape in Alna, Maine built in 1797.

Sills joined using Blind Mortise

In the basic New England cape, the sills were frequently joined at the corners using the Blind Mortise. I have also found it useful for joining side sills to sill girder beams.

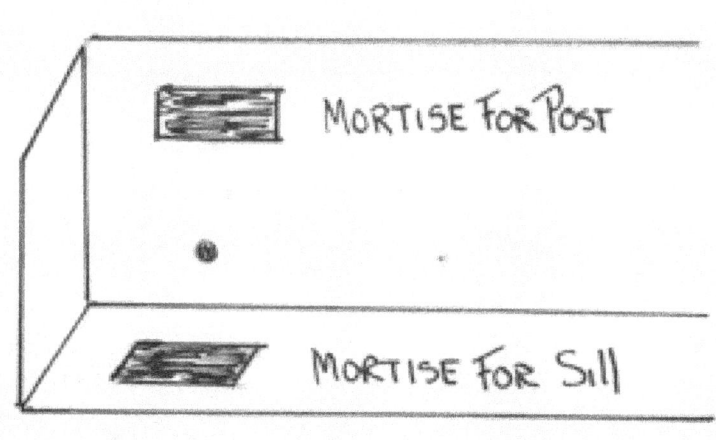

Blind Mortise

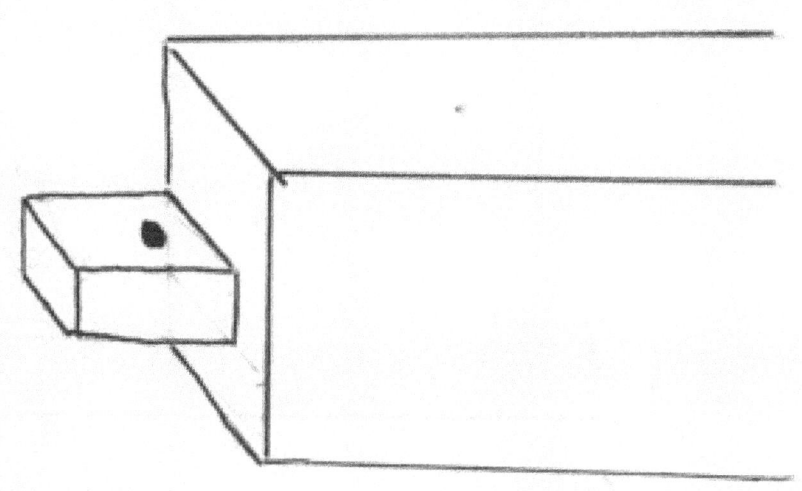

Sill Tenon recessed 2" from outside edge.
Layout for sill (view from top of timber): Mortise for sill is on inside facing surface 2" from bottom and 2" from outside

edge. Mortise is 2" wide by 3" or 4" deep. The length is determined by the width of the adjacent sill. If the mortise is 3" deep, the peg hole should be bored 1 3/8" from the inside edge. If the mortise is 4" deep, the boring can be 1 ½" from the inside edge.

Top Mortise is for post. Often the mortise is only 2" deep but will be deeper if the post tenon is to be trunneled. The mortise is 2" from each of the two outside edges and it is 2" wide.

The adjacent sill is tenoned. The tenon is 2" thick, 2" from the bottom and 2" from the outside edge. The length is 1/8" shorter than the mortise it is to fit.

The blind mortise works best when the sills rest directly upon a foundation. If the structure rest on piers, I modify the Blind Mortise to incorporate a shoulder to support the tenoned sill.

The sill girders are large beams spanning the width of the house. The sill girder carries the weight of the floor and often supports internal posts. The sill girder itself should be supported with posts, piers or foundation.

A replacement sill girder being slid into place . Tenon is 2" from bottom, full width of the timber and 4" long.

The sill girder generally is joined to the sill in the same way as the sill-to-sill corner is joined. The sill girder can either pass through the sills or butt against the sills. If the girder passes through the sills, the girder is blind mortised to receive the sills. If the girder butts against the sills, the sill is mortised to received the girder. The girder is tenoned. Not being near the edge of a timber, the tenon can be full width.

To make sure the sill girder is being supported by the foundation wall, I sometimes house the sill girder into the sill an inch.

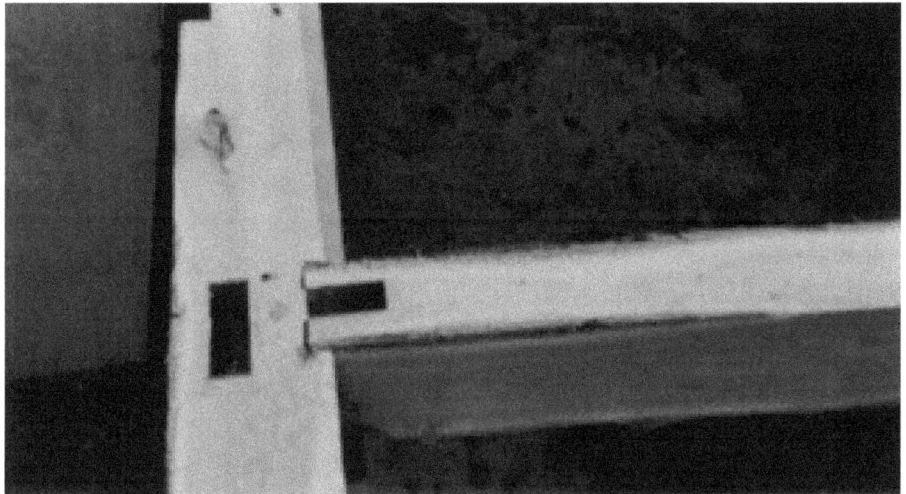

Sill girder housed to insure girder beam rest securely on foundation wall.

This is the common way to join sills and girders, but there are other ways and in certain situations these may be more feasible. The simplest solution for joining sill corners is the basic half-lap. Girders can also be joined to sills with the half-lap. Seeing as to how the floor joists will be supported by the sill girder, the girder comprises the bottom half of the lap. In this way it sits directly on the foundation or piers.

The method of joining sill corners that I prefer to use, especially when the structure will not be sitting on a continuous foundation, is the *bridled* sill. This is a very effective way of joining sills and corner post in structures that are to sit upon piers.

Sills lap joined   *Girder and sills lapped*

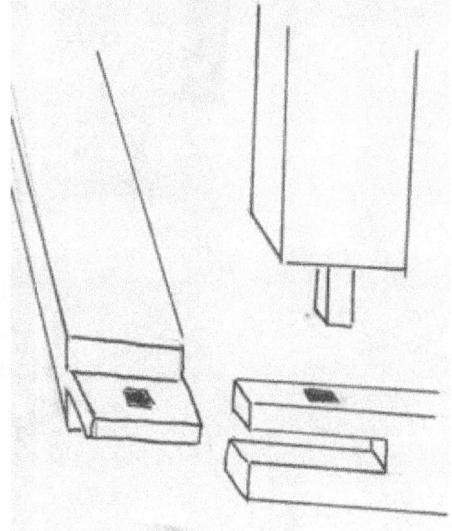

Bridled Sill

On this particular Bridled Sill pictured above, both sills and the post were 8" x 8". Measurements for the forks and tongue are as follows (from top to bottom): 2 ½" top fork, 2 ½" tongue, 3" bottom fork. The two-inch post tenon and the sill mortises were centered. The overall length of the tenon was 4 7/8". Once the sill tongue and fork were fitted and the post tenon was dropped in, the corner was very stable. The tenon, penetrating both the top fork and the tongue, acts as a trunnel holding the joint together. This is an excellent way to join sills and post especially when piers rather than a continuous foundation are used.

18th century capes had massive central fireplaces. In addition to the two continuous sill girders, chimney girts connected the two girders. The chimney girts on the cape whose floor system is depicted above were tied to the girder using a shouldered half-dovetail.

Chimney girt in restored 1797 cape

A shouldered half-dovetail joint

The difference between half and full dovetailed is the half-dovetail fans in only one direction while the full dovetail fans in both directions. Large floor girts can just as easily be joined using a full dovetail. If the floor system employs a summer beam, the summer beam is generally joined to sill and girder using a shouldered full dovetail joint.

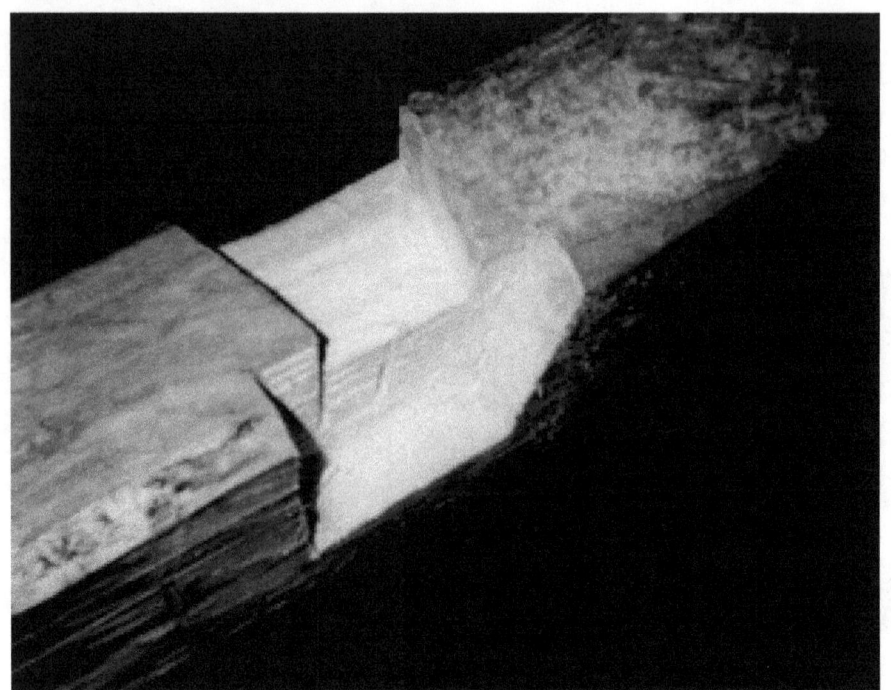

Mortise for a shouldered dovetail. The dovetail has a depth of two inches. The shoulder is one inch thick.

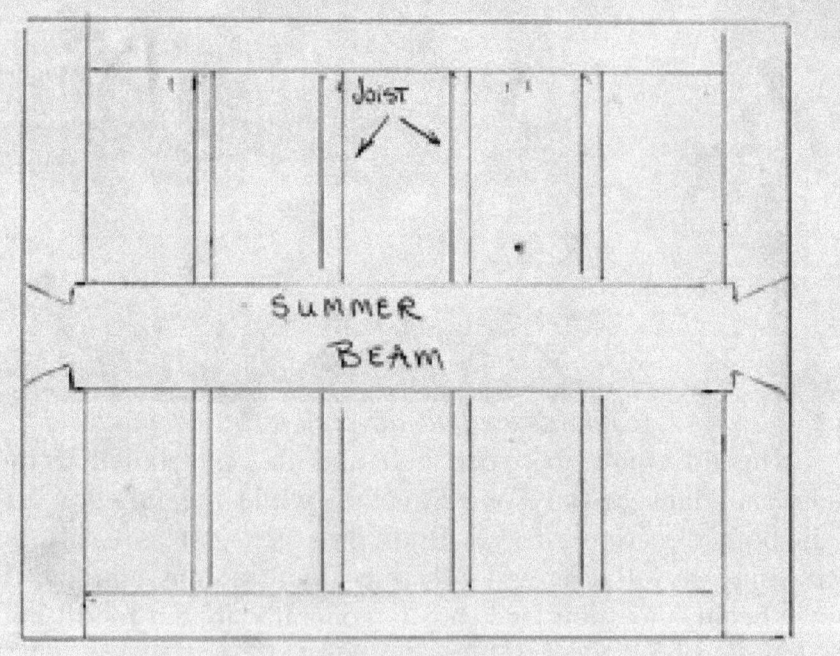

Floor system using a dovetailed summer beam

In log construction, sills present a unique circumstance. The sills do not lie in the same plane. Logs are placed one atop the other and at the corners the rise from one sill to the other must be half the height of the log.

In log construction, sills do not lie on same plane.

On this log house, sill seal and a pressure treated sill were set down first to create a barrier from moisture wicking up through the concrete wall. The log on the left was ripped in half with a chainsaw then planned smooth with the electric planner.

If the log structure is small and merely rest on corner piers this is no problem. If there are additional piers, the piers can accommodate this change of height. If, however, the structure is to sit on a slab or a full foundation, the practice today is to rip a log down the middle using a chainsaw. The two halves are then used as the bottom sills.

The half-log allows a seamless transition to the corner joinery.

Floor Joists:

Between the sill and the girding beams run the floor joists. Cutting joinery for floor joists is as basic as it gets. In the carrying timber one cuts a mortise two inches deep, as wide as the floor joists and as deep as the joists provided at least 2" to 3" of wood are left beneath the joist on the carrying timber. If not, the underside of the joists is notched.

Joist pocket is laid out onto carrying timber. The lines forming the outside edges of the joist pocket are cut with circular saw. The circular saw is set to a depth of 2". Joist penetrating 2" into a carrying timber is common.

The bottom of the joist pocket is bored out to a 2" depth either using a chain mortiser, or drill and bit. Once bored, the pocket is repeatedly kerfed with the circular saw to make removal of the waste wood with chisel and mallet easier.

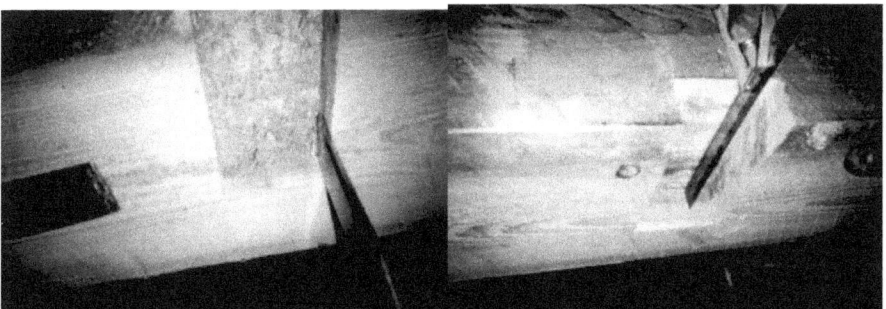

Joist pocket chiseled clean and checked for depth, length, width, plumb and square.

The joists are simply placed into the pockets, often tapped in with the mallet. If the joist is too tight, one simply pares the sides of the joist with a chisel or slick. Nothing good is accomplished by trying to beat an oversized joist into place. The joist are neither trunneled nor spiked. Gravity keeps them in place.

Joist undercut   A fitted joist

It was fairly common practice in early New England to leave floor joist in the round. The top surface was flattened to give a nailing surface for the floorboards and the ends of the joists were shaped to fit the joist pocket.

Joist left in the round. Top is flattened with adze or broad axe, ends are squared to drop into joist pockets.

Timber framers, in general, appear to have a fear of buildings spreading and joints failing due to this spreading. Frequently one or two joists per bay are tied to the carrying timbers. Long deck screws or common spikes will do the job in a pinch. The most effective method, however, is to use joinery.

I have used three different methods for *locking* joist. One method is to add a tenon to the joist and cut a matching mortise into the joist pocket of the carrying timber.

A tenon has been added to this joist. Once trunneled this joist will help resist possibility of spread in the structure.

Though drop in joists can wait until sills and girders are in place, tenoned joists must be fitted between sills and girders as the sills and girders are placed.

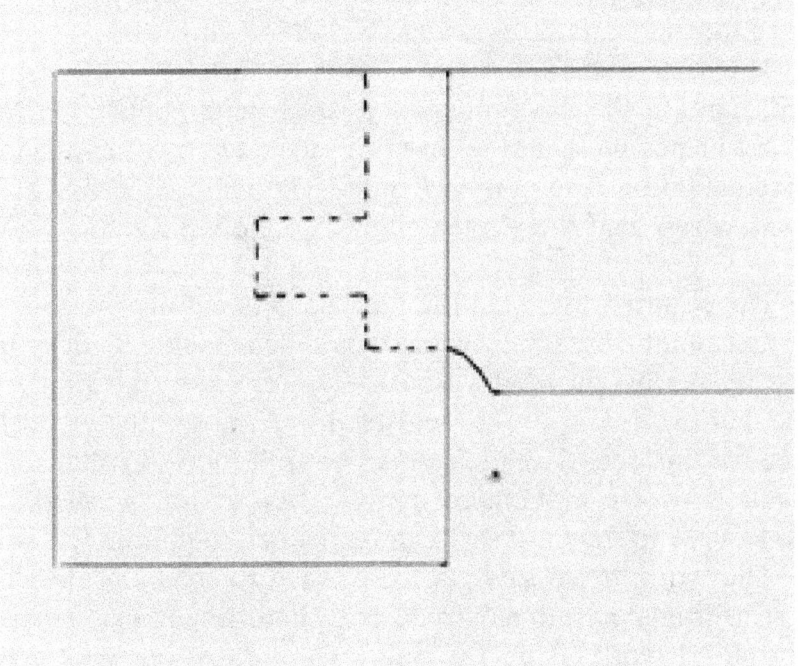

Cross section of tenoned joist into girding beam.

The other two methods I use are drop in methods. These *locking* joists can be put in at the same time as the regular joists. One is the shouldered dovetail, the same as depicted above for chimney girts.

Locking joists set in using shouldered dovetails.

The last method though similar is simply a full depth dovetail. Cutting the dovetail on a joist is quite simple. If the joist is six inches wide, at the point of intersection with carrying timber reduce it one inch on each side. The one-inch crosscut is done with a saw and the sloped angle is done with a chisel and mallet.

The easiest way to cut the mortise is to cut out the square pocket first. In the case below that would be a pocket 2" deep by 4" wide by the height of the floor joist. The dovetail line can then be cut with a saw and chiseled. A pull saw is fairly effective for this cut. The ideal situation, if a lot of dovetails are to be cut, is to use a standard circular saw and a worm drive circular saw. The standard saw can be pitched to cut the left side bevel. The worm drive saw can be pitched to cut the right side bevel. Unfortunately, no saw pitches in both directions. Then it is again chisel work. I do the bulk of the clean out with a 2" framing chisel, but to get into the tight corners, I change to a one inch chisel.

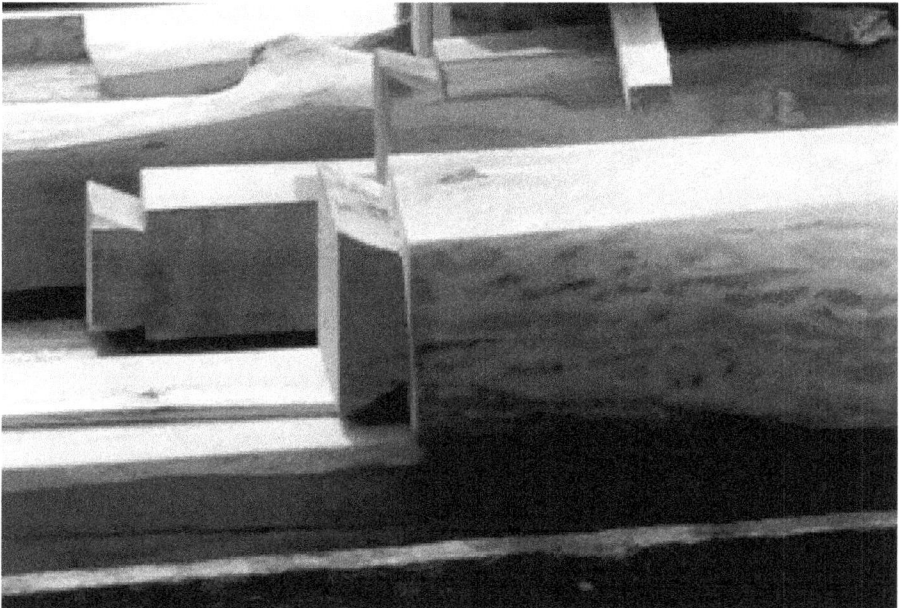

Full depth dovetailed joists

Fitted dovetailed joist

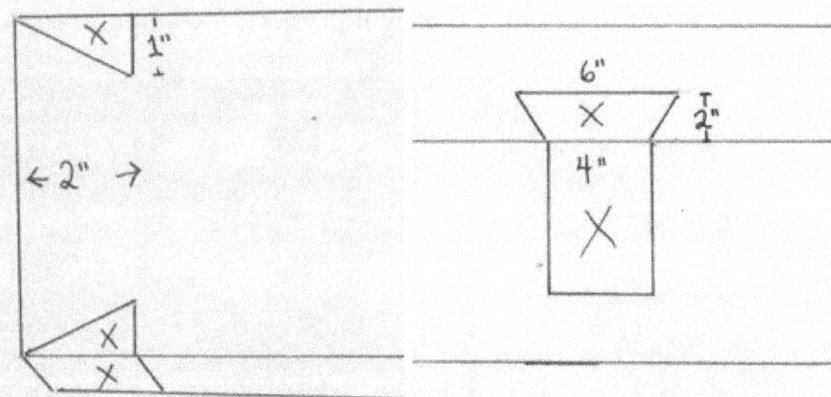

Layout of dovetail floor joist and accompanying mortise in carrying timber.

  The top layout for the carrying timber is the inverse of the joist. 2" into the carrying timber is the straight line, the width of the timber. At the edge of the carrying timber the measurement comes in one inch on each side. The two points are connected. These two lines create the dovetail. Down the side of the timber it is simply two straight lines drawn from the narrowest part of the dovetail.

Chiseling out a full dovetail joist mortise

Full depth dovetail mortise.

If all the floor joists are dovetailed, it creates a very strong, rigid structure.

I usually place floor joists 24" on center for the first floor and 32" on center for the second floor. With kiln dried 2" V-match decking it is possible to place the joists 48" on center.

Occasionally a structure presents a challenge that conventional framing cannot answer. The plan on the structure below called for the joists to be below the tie beam and to protrude through the wall. These joists are notched to the underside of the tie beam and bolted. Rather than resting on the tie beam, they securely hang from it.

Joists on this structure hang from tie beam with carriage bolts.

The second floor joist system is one of the most noticeable aspects of a structure. Regardless of the style or size of the structure, one should make an effort to make it aesthetically pleasing.

Bracing

Without bracing a timber frame will rack and eventually collapse. The massive timbers in a frame guarantee the frame will be capable of supporting immense weight. It is, however, the small members of the frame, the braces, which give the frame its rigidity. The average frame stands unflinching because of 3"x5" timbers less than 5' long.

Braces work in conjunction with one another. They oppose one another. A basic mortise and tenon brace works in compression, meaning that it absorbs forces that compress the shoulders of the brace against both horizontal and vertical timbers. Mortise and tenon braces, however, do not work well in tension, meaning the forces that seek to pull the brace tenon from the timber mortise.

Braces are one of the most numerous components in any frame. A basic 22' x 36' frame can easily have as many as 30 braces. Unless the braces are to be scribe fitted to the frame, braces generally are interchangeable and fairly easy to cut.

← ← ← ←   → → → →

  Mortise and tenon braces absorb forces in the direction shown. The braces absorb the forces moving toward them not away from them. These are the forces of compression. To absorb forces in each direction requires two opposing braces.

  3"x5" stock is the most common brace stock size used. I have also seen 4"x5", 3"x6", and 4"x6" to name a few. I generally use 4"x6" stock. The braces shown in the above two photos are 4"x6". The reasons I choose 4"x6" timber are three:

  If the timber is to be hewn from a log, it is easier to keep steady and hew than it would be to hew out a 3"x5" brace. The 3"x5" would bounce around on the cribbing and have a tendency to shake off the log dogs.

  If sawn stock is to be purchased, 4"x6" stock is standard and usually in stock, whereas 3"x5" is generally a custom order.

  4"x6" braces require a 2" mortise. 3"x5" braces require a 1 ½" mortise. I do all my post/plate/tie beam mortises 2" wide. Doing my brace mortises the same width keeps me from having to change my 2" self-feed bit to a 1 ½" bit.

  The measurements given here, therefore, are for braces using 4"x6" timbers.

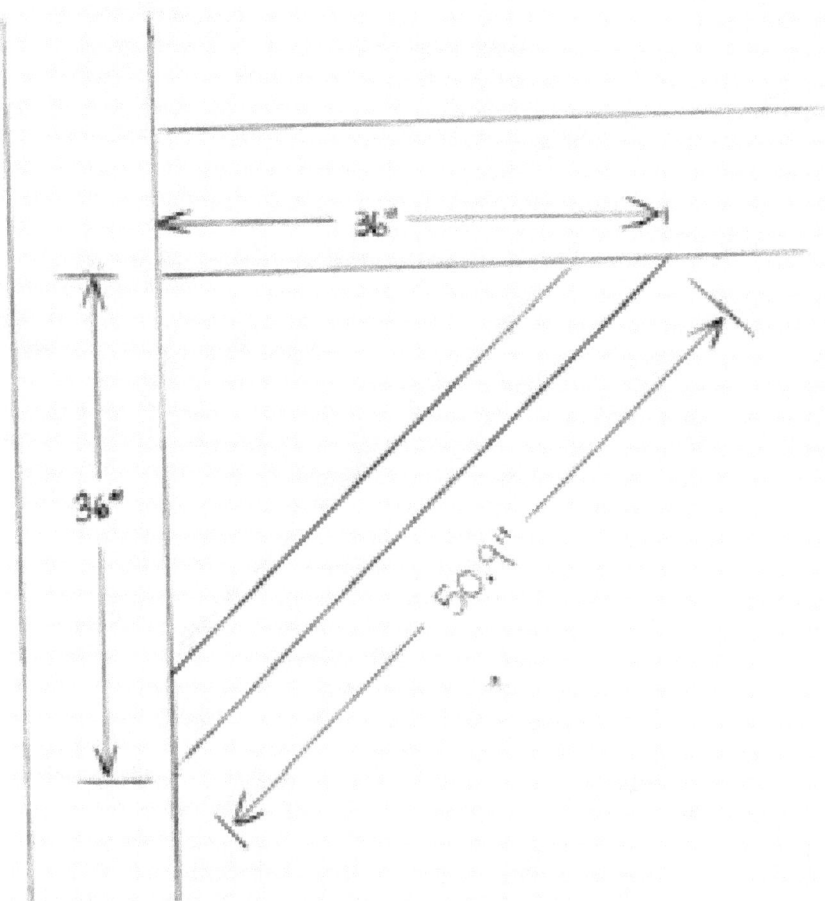

Braces are generally set equidistant from post and girt, plate or tie beam. A common distance is 36". In small frames that figure can drop to 30". In large barns that figure can rise to as much as 48" or 60". The length of the brace from shoulder to shoulder is determined by the Pythagorean formula: $a^2 + b^2 = c^2$, which comes to 50.9" for a brace set 36" from intersection of two timbers, 42.4" for a brace set 30" from intersection of timbers, and 67.9" for a brace set 48" from intersection of timbers.

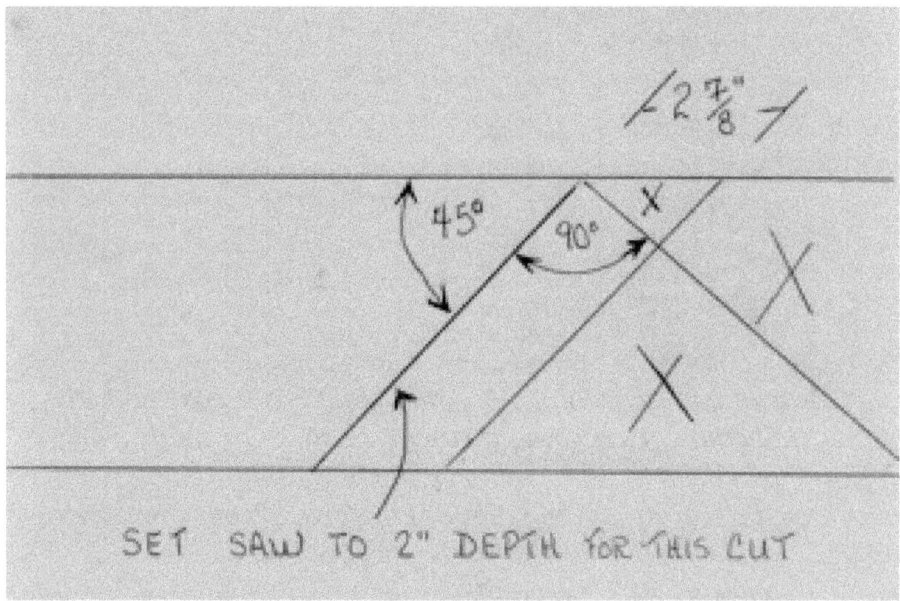

Basic layout for brace.
--*The areas marked with an X are to be removed.*
--*The line shown with arrow must be cut to a depth of 2".*
--*The layout is done on the outside facing surface (the reference face).*
--*The brace is then tipped on edge. The framing square is aligned with the outside edge (reference face) and the line scribed. (photo below)*

--After scribing line, a rip cut is made with circular saw.

--Brace is flipped back, chisel is placed into 2" saw cut and struck. Waste chunk should fall off. Some cleanup with chisel may be necessary and the edges should be chamfered.
--Process is repeated for the other end.

As mentioned in the chapter *Mortise and Tenon*, a 4"x6" timber may not in fact be 4"x6". If after cutting, the tenon is over 2" thick, it should be pared to 2" on the *non-reference* face. This keeps the finished brace sitting flush with the outside of the building.

Mortise layout: Mortise is laid out 36" from the intersection of the timber it is to brace and 2" from the reference face. (If on an outside wall, reference face is the side that faces out.) The brace mortise is 2" wide and 8 ½" long. I define the bottom 5 ½" differently than the top 3", which I label slope. If using a self-feed bit, a centerline is useful for positioning self-

feed bti. The three Xs define actual placement of screw tip of bit for boring. The Xs are marked at 1", 2 34", and 4 ½".

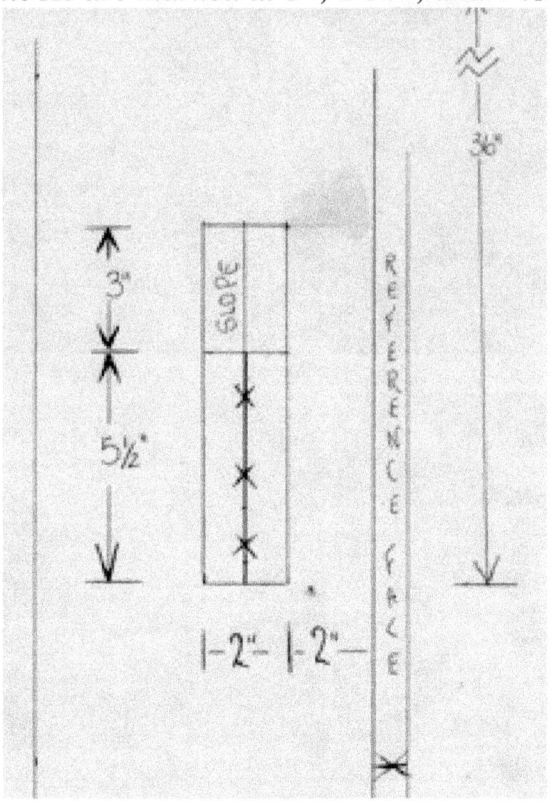

Brace mortise layout

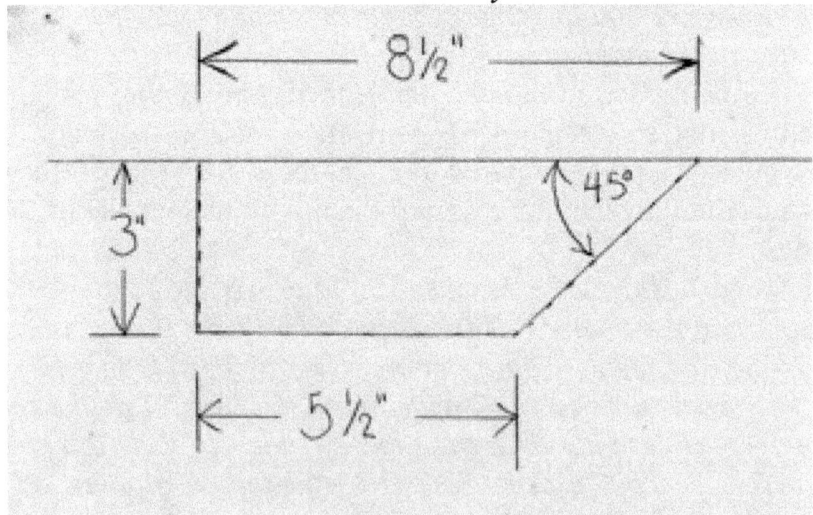

Cross section of brace mortise into timber

The brace system described is a very common basic *mill rule* brace system. It is simple, easy to cut, and very strong. Frequently, however, people shoulder in the braces. This is a matter of course in *square rule* framing and quite common in general.

These braces are shouldered in this square rule frame. (Frame cut by Bud Menard)

Shouldered brace mortise laid out.

On this particular frame, shoulder depth is laid out to a depth of 1" (one inch less than the nominal dimension of the timber measured from the reference face. See chapter Mortise and Tenon.)

Plunge cuts are made full depth of saw along lines running parallel with beam (ripcuts).

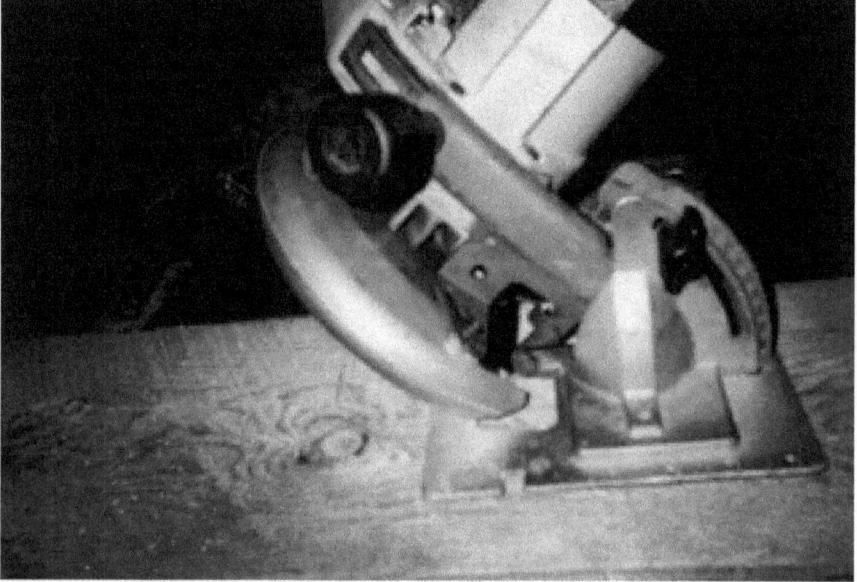

The beveled and cross cuts are made to the depth of the shoulder

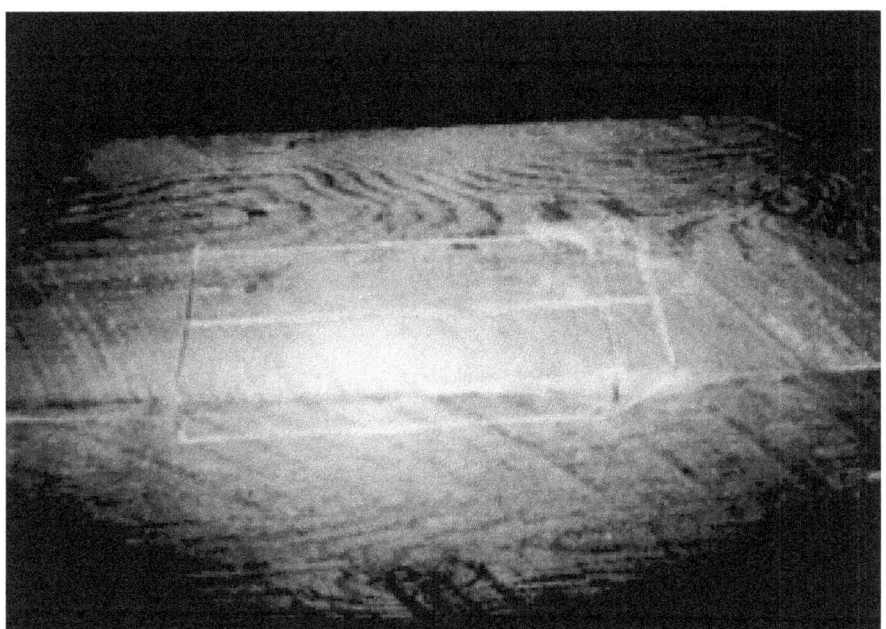

. Once all saw cuts are made, the shouldered mortise is ready for boring and chiseling.

Mortise is bored to a depth of 4" (3" for tenon plus 1" for shoulder), then it is cleaned out with a chisel.

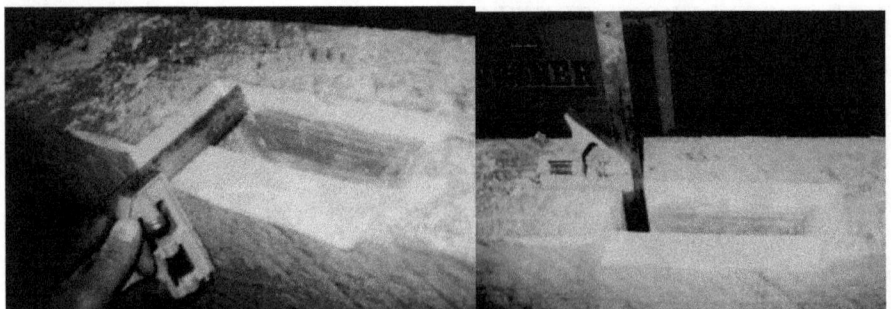

Mortise and housing must be checked for level, depth, width and plumb.

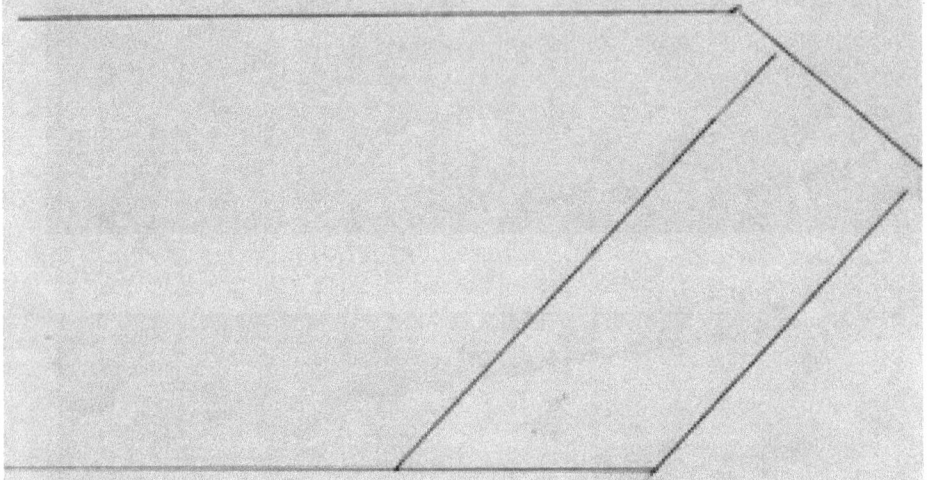

If the brace mortise is housed, the brace is likewise modified.

Of course, not all braces are mortised and tenoned. Braces may just as easily be lap-joined to the timber from the outside. This is particularly advantageous when the brace is to be scribed.

When scribe-fitting braces it is possible to raise the entire frame without braces. The frame can be racked and plumbed and temporary bracing nailed upon it to hold the frame steady. One can then go back and scribe the braces to fit.

When I first started framing, scribing several of the braces in after raising the frame was common practice for me.

I have on occasions tried to pre-cut brace laps and the mortise pockets on hewn timbers in a generic fashion, assuming the braces would be interchangeable. The results were far from desirable and I would not recommend anyone doing it.

Braces scribe fitted and lap-joined

One person holds the brace steady as the other transfers scribe lines.

Though today I tend to mortise and tenon my braces in the manner described earlier, there are still many times where I find it to my advantage to scribe the braces in after raising. With curved, natural forms, scribing is the simplest way to fit the timbers.

Curved and wavy apple braces scribe-fitted.

If one is going to lap fit a brace, one can cut a half-dovetail instead of simply a straight lap. Unlike the conventional mortise and tenon brace, such a brace works not only in compression but also in tension. Working in tension, such a brace can be used to hold a knee wall, for instance, from kicking out, where a conventional, compression style brace would have minimal effect.

Half-dovetailed brace works in both tension and compression. In this case the brace is helping to hold the knee wall from kicking out due to the outward thrust of the rafters.
(Brace fitted by Bill Rispoli)

    I am certain there are other options available for one bracing a frame. These techniques, however, basic mortise and tenon, shouldered, and scribed, lapped and dovetailed should meet ones every need in timber framing.

Close up of a scribe fit half-dovetail brace. Brace was fitted in after frame was up.

Tying it all together

Roof systems are heavy, and, with the snow and ice load winter piles on them, the roof becomes even heavier. As all this weight pushes down upon the roof rafters, that force is transferred as outward thrust to the post and plates. Included in the design of any structure must be something to counter or restrain that outward thrust. In timber framing, that something is called the *Tie Beam*. How much outward thrust the tie beam must counter depends upon the design, the size of the structure and the weight of the roof.

I was approached a few years back to build a small 18' by 24' hand hewn timber frame that would house a sauna and changing room. The man wanted eventually to sod the roof. He asked that the timber frame be built with a 1:12 pitch.

With such a shallow pitch, the easiest way to control outward thrust is simply to cancel out downward push by supporting rafters from beneath. On this particular structure the downward push of the rafters was absorbed by three large horizontal timbers resting on tie beams and supported by post.

In this frame, downward thrust of rafters is absorbed by three horizontal timbers (purlins) resting on tie beams.

In its most basic form, a tie beam is simply a beam running from wall post to wall post, parallel with the rafters and joined to the posts.

Though I seldom see it used in Timber frames, I have been exposed to one of the easiest ways to cancel out the downward push of the rafters. It is to rest the rafters upon a continuous ridge beam supported by braced King posts.

The braced King post of this structure absorbs downward push of rafters, greatly reducing need to tie against outward thrust.

Continuous Queen posts minimize need to tie against outward thrust by absorbing downward push of rafters. The Queen posts, tenoned directly to the rafters, transfer the weight of the roof to the concrete base rather than out to the end wall posts.

Thru-tenon tie beam.

Thru-tenon is 2" wide, 2" from outside of post.

As one moves away from interior post absorbing the downward push of the rafters, we must restrain that push where it will surface, which is outward thrust to the plates and exterior posts, and that requires a tie beam.

A very simple tie beam joint that works well on small structures that have no second floor is a thru-tenon. The tenon is 2" thick and 2" from the outside surface. It is as wide as the tie beam and as long as the post is thick. It is held by two staggered trunnels.

If the tie beam is to support a second floor, it is necessary to shoulder the tie beam into the post.

When using a straight tenon as a tying joint, we are relying to a great extent on the hold of the pegs in keeping the beams tied. By changing the joinery slightly, the strength of the joint can be greatly increased. Rather than a straight tenon passing through the post, one can use a half-dovetailed shouldered tenon.

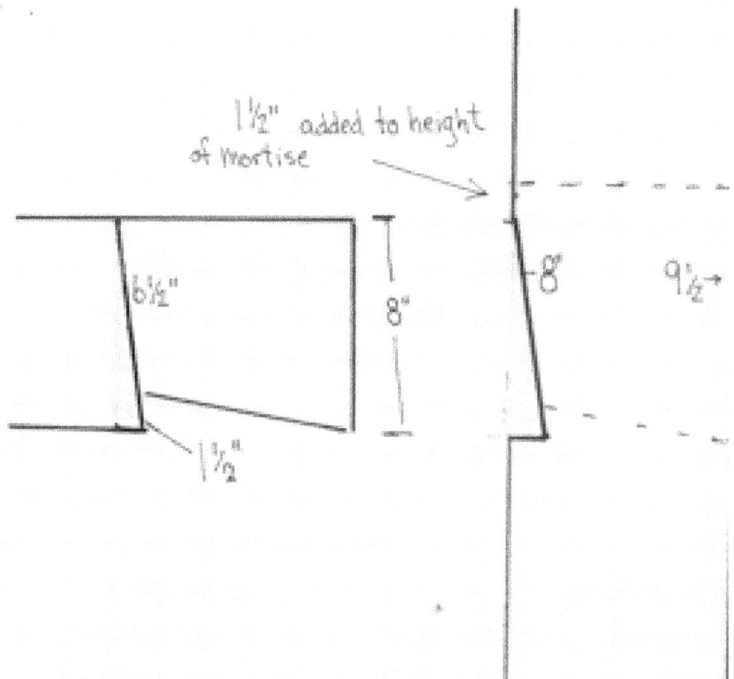

Layout of half-dovetailed tie beam joinery

In addition to a trunnel through the tenon, the dovetailed tenoned is held tight with a wedge. If the half-dovetailed tenon is recessed 1 ½", then 1 ½" is added to the height of the mortise. This allows the half-dovetailed tenon to slide in. It is then driven tight by a hardwood wedge. A good wedge would be slightly tapered and slightly more than 1 ½" on the wide end.

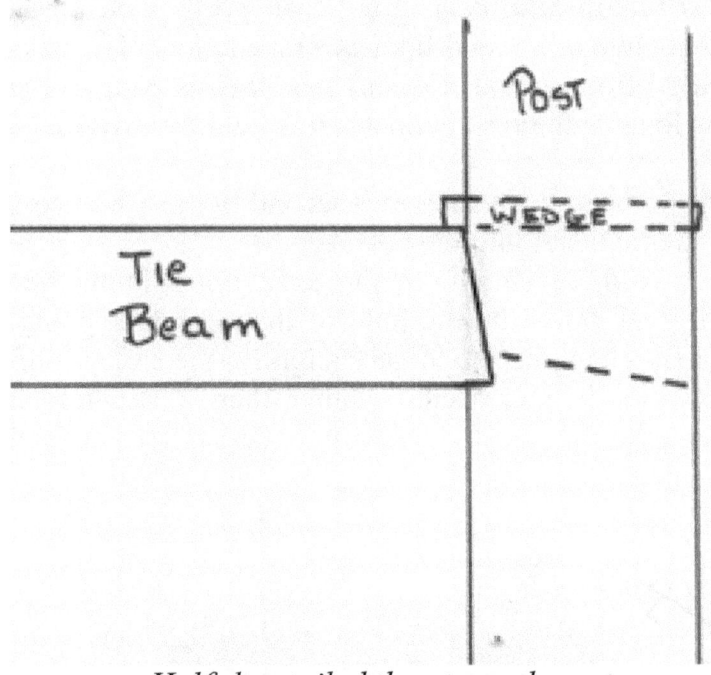

Half-dovetailed thru-tenon layout

Half-dovetailed tie beam tenon

Half-dovetailed tie beam fitted. Joint is shouldered, trunneled and wedged.

The through tenon and the half-dovetailed tenon are tie beam connections used where the tie beam joins to the post below the plate. Very different tying joints are used where the tie beams sit above the plates. Perhaps the most well known of these is the English Tying joint.

It is called the English Tying joint because this is the tying joint the English brought with them to the colonies. There are 17th century examples of houses joined in this way still standing in Massachusetts and Connecticut. It was the general way to build throughout the 18th century and continued in rural New England states such as Maine until the mid-nineteenth century.

The typical New England cape is usually framed with the English Tying joint, and the earliest barns were likewise. The joinery is so much apart of these barns that we generally refer to them as English barns. The current revival in Timber framing really began to flower in New England in the 1970s as people began to disassemble barns. It is no surprise the English Tying joint also made a revival.

The English tying joint is exceptional because it ties the post, plate, tie beam and rafter into a single unit.

The post top is two tiered with a lower shelf for the plate and the top shelf for the tie beam. Each shelf has a tenon for the respective beam. These tenons are perpendicular to one another, just as the plate and tie beam are.

English Tying Joint (photo above: with overhang at eaves, photo below: no overhang at eaves.)

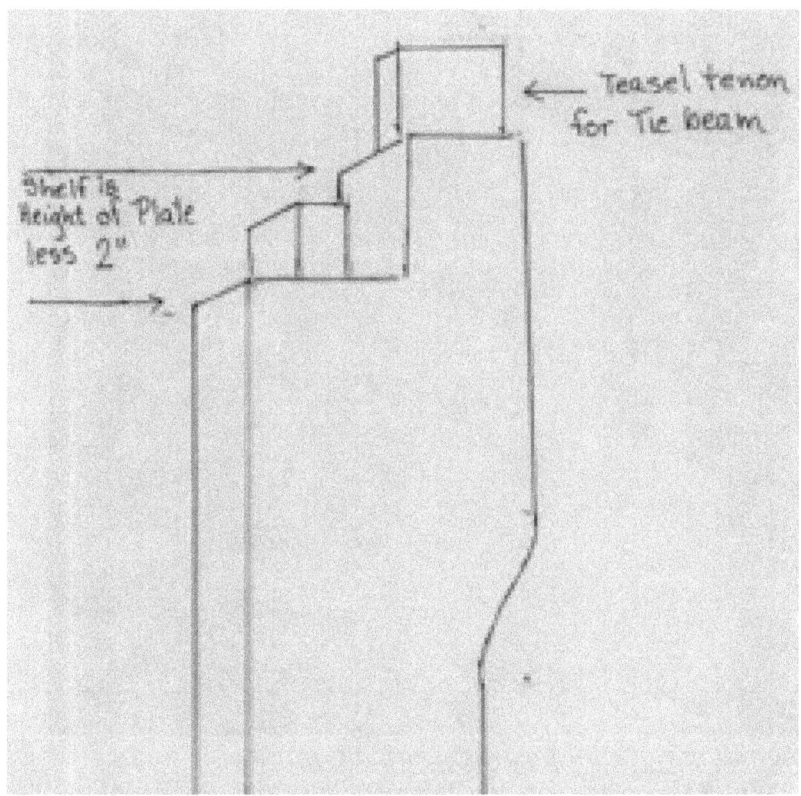
Layout of post top -- English Tying joint.

Post top – English tying joint

English tying joint post top utilizing natural flare of tree

When the plate sits on the post, the top of the plate rises 2" above the top of the post (excluding the teasel tenon.) To accomplish this, the shelf height is cut 2" less than the height of the plate. The width of the shelf is generally the width of the plate. If that is not possible, the inside surface of the plate will have to be relieved to fit.

The tenons are two inches thick and either centered or 2" from the reference face. The post is often flared. Eric Sloane referred to it as a *Gunstock* post. The English call this a *Jowled* post.

Into the top two inches of the plate is cut a full dovetail. (If this is a corner of the building and the plate does not extend beyond the post, I cut a half-dovetail.) The dovetail is the width of the tie beam at the wide end, narrowing down 1 ½" or so on each side.

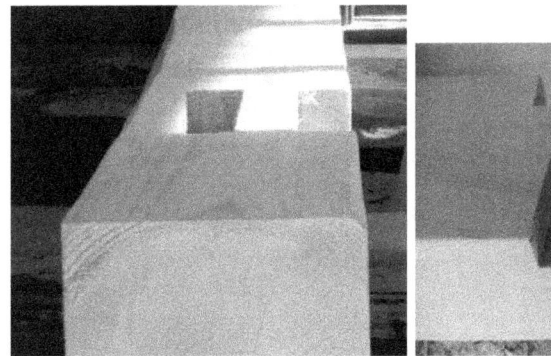 

Left -- The underside of the plate is mortised to accommodate the tenon. This plate is 8" wide, the post shelf for the plate is 6". The 2" relief to the right allows the plate to seat down onto the post. Right -- A 2" deep full dovetail cut into top of plate.

The layout for the Dovetail on the Tie beam is easy. The Dovetail is a match of the opening cut into the plate. The length of the dovetail, the amount that the dovetail recedes, and the depth of the dovetail are all the same measurements we used on the plate it intersects.

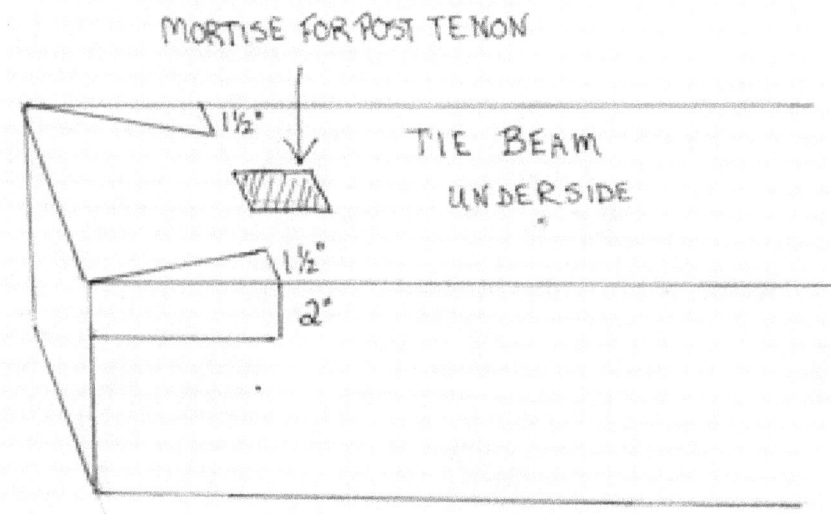

Underside of Tie Beam.

The angled lines on the underside can be cut with a circular saw. Simply set the depth and make the cuts. The stop cuts can be partially cut with a hand saw. The rest of the work is all chisel work. The depth line needs to be scored with chisel and mallet. The stop cut also needs to be finished with chisel. One should be able to pop the wedge out and do a final clean up with the chisel.

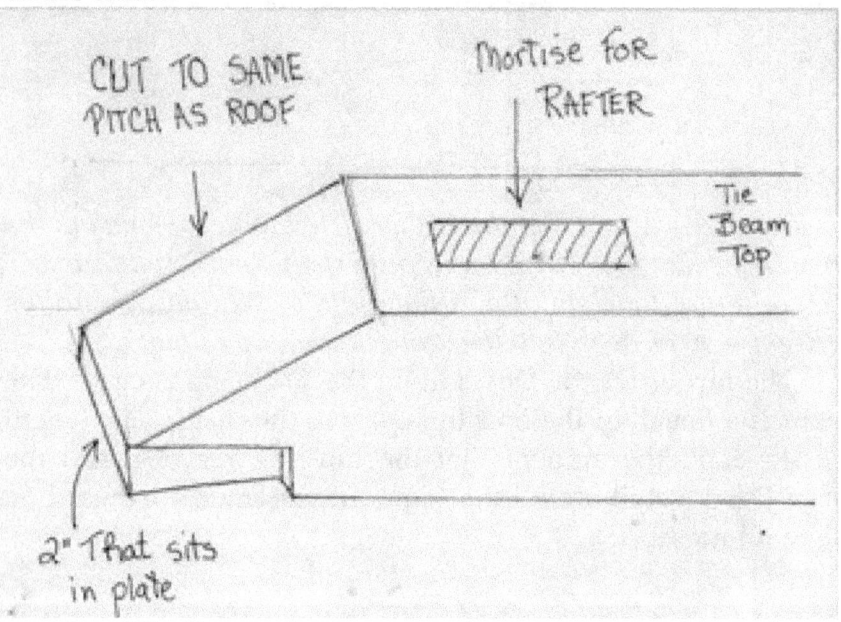

Topside of Tie Beam. (with no overhang at eaves. If overhang is desired, extend tie beam out beyond dovetail at plate.)

    The mortise for the post tenon must be bored and chiseled clean. Distance from the reference face is the same as the distance from reference face used to cut tenon on post. (illustration on p. 153)

    The top of the tie beam ties the roof rafter to the tie beam. The tie beam is cut to match the roof pitch. The last step is to cut a mortise to accommodate the rafter. If the rafter is six inches wide, I cut a 2" wide mortise, 2" from the reference face. If the rafter is eight inches wide, I cut a 2" wide mortise, 3" from the reference face. The mortise should be either 3" or 4" deep. Anything less, there would not be enough tenon to peg. The back end of the mortise can be sloped to match the rafter pitch. An 8" x 8" rafter with a 10:12 pitch would have a 9 5/8" base. If the first four inches were left without a tenon, the tenon would be 5 5/8" long. The mortise in the Tie beam would also be 5 5/8"

155

long. The back side of the mortise would be sloped to match the rafter pitch.

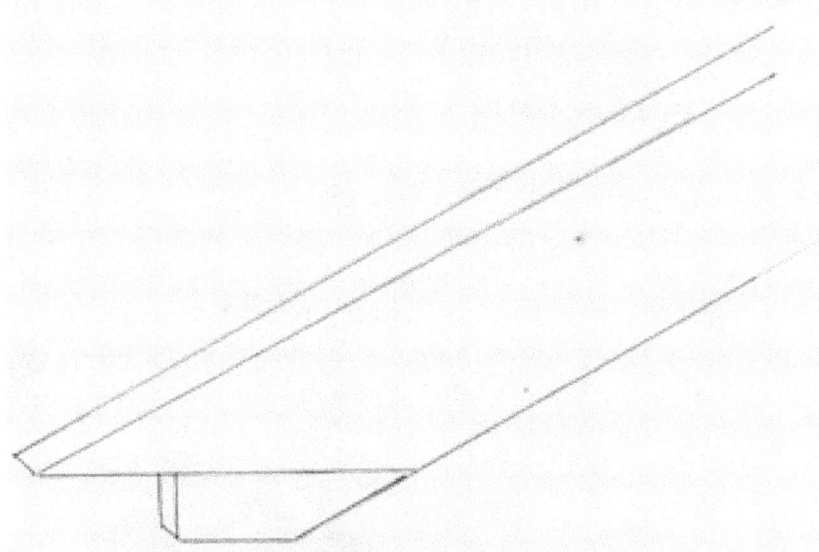

Tenoned Rafter in English Tying joint.

Here we see all the elements of the English Tying Joint.

The English Tying joint is like an icon. It was used extensively in early New England homes and barns, but not always as depicted above. The English Tying joint embodies an ideal that quite often was modified. These modifications all have specific features in common.
1. The Plate is below the tie beam and joined to the post with mortise and tenon.
2. The tie beam is joined directly to the plate.

I have had the opportunity to work with two such modifications. One was a very simple modification on a late 19th century Queen post truss barn. A simple relief was cut into the bottom of the tie beam allowing it to fit snugly over the plate. The post was also modified in that a straight post was used with but a single tenon for the plate.

Modification of the English Tying Joint on a late nineteenth century Queen trussed barn. The relief in the tie beam fits over the plate. A single trunnel holds the tie beam to the plate. This simple design requires no joinery be cut into plate top.

A cog instead of a dovetail can be used to anchor the tie beam to the plate. The cog is the projection on the bottom right of the tie beam. It drops into a matching mortise on the plate.

In Spring of 2006 I was asked to restore the frame on a 200 year old barn and to simply replace the missing timbers with sawn Hemlock beams. When I arrived at the site and saw the stack of timbers, I noticed the flared post and immediately assumed English Tying joint. When I looked closely, however, I was totally perplexed. Though the hewn posts were visibly flared, the flare served no purpose. There was simply a tenon on top of each post. The tie beams were missing, but it was clear the tie beams were held to the post with a simple mortise and tenon connection. The plates were intact, and were tenoned to fit into mortises into the side of the post like girts. As I looked at the puzzle, the roof structure became apparent to me. This frame was a *square rule* frame but the *scribe rule* habit of flaring the post for the English tying joint was probably conditioned habit for these builders. If one is looking for an easy way to tie a structure, this post top connection of the tie beam with the plates acting as girts is fairly hard to beat.

158

Simple mortise and tenon tying post to tie beam.

It is, however, designed for a *Purlin* roof system. (See Roof Systems). The plate is set too low for a Common Rafter system.

Girding Beam:

In a full two story building, in addition to the tie beam, we have additional timbers that span the building. There are principal timbers that carry the second floor joists. These timbers are called *Girding beams* and they generally (though not always) run parallel to the tie beams. These are sizeable timbers, like the sill girders beneath. These timbers carry considerable weight. They are generally supported along their length by interior post.

Since the tie beam restrains against outward thrust, the girding beams are concerned primarily with the downward push of the weight of the second floor. To prevent the tenon shearing off the girding beam due to this downward pressure, the girding beam is commonly thru-tenoned and shouldered into the exterior post of the frame.

One full floor beneath tie beam is the girding beam. Girding beam is thru-tenoned and shouldered into corner post.

The tenon is the full height of the girding beam. Two trunnels are used at each post to tie the girding beam. Unlike the tie beam, it is neither dovetailed nor wedged.

Hammer beam truss (built by Bob Brann for Windsor Fair Grounds, Windsor, Maine)

Of course, since Medieval times craftsmen have found ways to deny the need for tie beams. The Hammer Beam Truss permits an open space from floor to roof rafters.

The outward thrust of the rafters is not limited to timber frames. It is an issue for log buildings as well.

In Appalachian and Quebecois style hewn log construction where the gable ends are not log, a simple common solution to resisting the rafters' outward thrust is to secure the plate at the corner notch with a long 1 ½" locust or oak trunnel or a 1" steel pin. This pin must protrude through several logs.[5]

---

[5] Langsner, Drew, *A Logbuilder's Handbook*, Rodale Press : Emmaus, PA, 1982, p. 161.

Hewn log house in Quebec

Scandinavian log builders developed a solution that transfers roof weight directly to the solid log, gable end walls. Principal purlins are built with locking joinery directly into the log gable walls. In such a design, outward thrust is completely eliminated. Outward thrust of the roof must always be eliminated, neutralized or restrained.

Outward thrust of rafters is absorbed by gable walls in Scandinavian log building. (Norsk Laftskoale, 2007)

Mc Raven, Charles, *Building and Restoring the Hewn Log House,* Betterway Books, Cincinatti, OH, 1994, p. 102.

Roof Systems

As a framer, I find roof framing to be the most exciting part of the frame. It makes the structure look complete and brings out the character of the building. Roof framing can be quite simple or very complex. It can be totally practical or amazingly artistic.

Contemporary, conventional, light framing frequently uses a common rafter system. The common rafter system is actually a popular system in both timber framing and log construction. However, in log and timber construction a whole variety of roof framing systems are possible. There are several versions of beautiful, heavily framed truss systems. There are combinations of purlin (*horizontal roof framing members*) and rafter systems. Sometimes the purlins are the principal structure in the roof and the rafters play a secondary role. Other times the rafters are the principal structure and the purlins are secondary.

In the early 1970s as friends were leaving suburbia to live in tents, teepees, and quickly erected shacks in the rural woodlands, I became fascinated with the idea of purchasing a few country acres and building my own home, barn and outbuildings. I envisioned myself living a rustic life in a log cabin I built myself. Of course I had absolutely no experience in building. I began purchasing books, but I avoided any books that appeared too technical or detailed. One book I did purchase, among others, was Ben Hunt's *Building a Log Cabin*, written in 1947.[6]

Claudette and I bought the land in 1976 but circumstances delayed our move onto the land until Spring 1979. With logs I had hewn the year before and piers I had likewise built the year before, I commenced to build a log home. I looked at my books to the easiest possible approach to cutting the corner notches for hewn timbers. Ben Hunt's illustration of the Square Notch looked the simplest. (Ben Hunt mistakenly called this notch a Dovetail Notch.) Of course I did not hew enough logs and more had to be hewn. And there were other interruptions. It all progressed slowly but the walls were finally up come September. Floor joists were muscled up to the second floor and spiked to the plates and central girding beam.

It was now time to put the rafters up and I realized I did not have a clue how to proceed. I knew that if the roof peak was square, the angle at the plate was forty-five degrees. And if the cabin was 18' wide, the roof height would be 9' tall. That, however, was all I knew. Pythagoras may have given me the formula in geometry class, but it somehow never occurred to me in any practical sense that this information could have helped me actually figure the rafter lengths.

I looked at my books. Ben Hunt used a ridge board on his cabin, so did Roy Thompson in *The Foxfire Book*. My brother came to help. We knew the height the ridge should be. It looked like a simple approach: two forty five degree cuts and nail the rafter into place.

We temporarily braced the ridge board into place. We put some temporary decking boards up and placed the rafters upon this temporary second floor. A measuring tape strung from the top of the ridge board to the edge of the plate gave us the length.

---

[6] Hunt, W. Ben, *How to Build and Furnish A Log Cabin*, Collier McMillan Publishers : London, 1974, p.46

I wasn't sure I could get a straight cut with the chainsaw. My brother Den looked at me and said "Think your way straight through." This Zen statement was all I needed. Soon we had the rafters up and removed the temporary braces from the ridge pole.

Rafters butted against a ridge board and spiked.

The roof was boarded with rough sawn one-inch boards and covered with 90 lb. rolled roofing, which unfortunately has a short life.

I was not concerned with overhang because I intended to add on to the house as soon as I could afford too.

The gables were boarded and shingled. Windows were put in -- And then the Winter winds came.

In contemporary carpentry they use plywood or Advantec, a form of OSB. Although plywood appears very weak, unable to support its own weight, once nailed it is very strong. Unlike boards, plywood not only sheaths but acts as wind bracing because each nailed sheet is 4' wide. I used low grade, rough cut boards to sheath my roof and I had never heard of wind bracing.

Whenever a severe windy storm occurred, I thought my roof was going to collapse. I knew something was wrong but I still hadn't figured out what I had done wrong. A couple of years later I added on to the rear section of the house. This addition involved a gable projecting off the roof system and the problem disappeared.

Gabled addition to roof acted as wind bracing.

The gabled addition provided the diagonal bracing the roof needed to stabilize it in extreme windy conditions. One gable dormer can stabilize an entire roof. If boards are used, particularly in a windy situation, wind bracing ought to be considered. If wind bracing is included in the first and last rafter set or a gabled dormer is added to the roof, there is no reason why a simple roof with logs and a ridge board cannot work. Thirty years later mine is still standing.

Despite what I have stated about wind bracing. A year after building my log cabin, prior to the gable addition, I helped Bill Behrens build his barn. (See chapter entitled Raising.) As with my cabin, the pole rafter feet were simply cut to sit flat on the plate. The tops were half-lapped. The rafters' butts were pegged to the plates with one inch pegs as were the peak laps. Every other rafter had a collar tie.

Raising rafters on Bill's barn

Two by fours were nailed perpendicular to the rafters to receive the metal roofing. No diagonal bracing was ever put into the roof system, but between the oak pegging, the two by four nailers, and the metal roofing, the wind pressure must have been neutralized because no wind bracing was ever needed. The barn has stood since 1980 and still stands true today with no noticeable problems.

In both 2006 and 2007 I built common rafter frames, both 16' wide, one 48'long, the other 22'long. The 4" x 6" rafters were set 32" on center. On one, instead of structural insulated panel sheathing, 2" x 4"s were nailed every 24" for attaching metal roofing.

The nice thing about a common rafter roof is that after a rafter has been calculated and cut, this rafter can serve as a template for the rest.

When calculating rafters we usually refer to roof pitch as a ratio of inches in height per foot run. In an ordinary house, the peak of the house is plumb with the middle of the width of the house. The horizontal run is half the width of the house. If the structure is twenty feet wide, the run is 10'. If we say the house has a 9:12 pitch, we are saying that the height of the roof rises 9" for every foot of run. In this case the peak of the roof is 90" high (9" multiplied by 10).

A common rafter roof being strapped for metal roofing

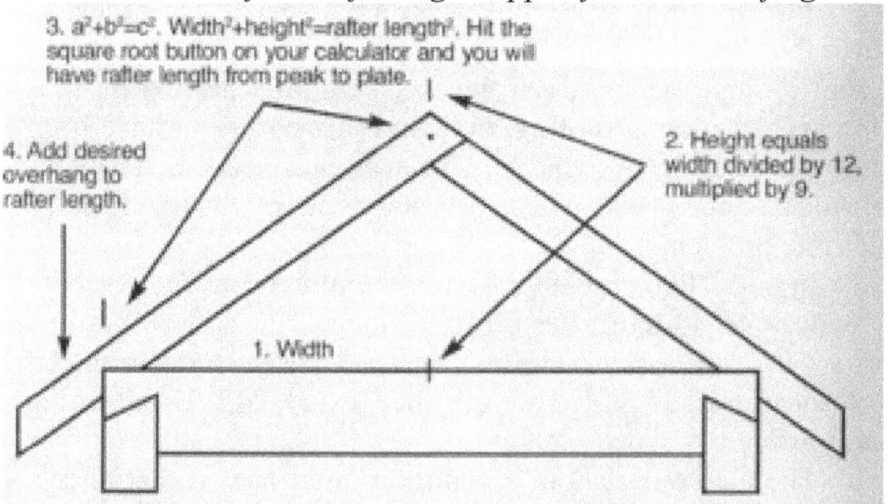

To determine rafter length, we use Pythagorus formula – $a^2 + b^2 = c^2$. A = the run, 120". B = the height, 90". Plugging these numbers into a calculator, we find $C = 150$". To this figure we then add whatever overhang we want.

Roof pitch as a ration of inches rise to foot run also makes it very simple for us to determine the plumb and horizontal angles the rafter peak and tail will have to be cut. By aligning the framing square at 9" and 12" onto the edge of the timber, we get both the plumb and the level cut.

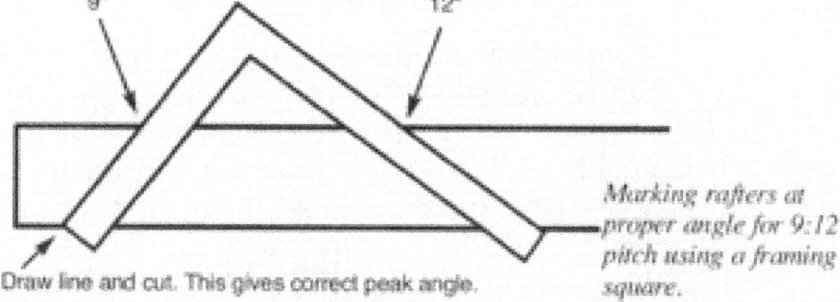

Marking rafters at proper angle for 9:12 pitch using a framing square.

Because 9" represents the rise, the arm of the framing square set at 9" gives us the plumb cut. And because 12" represents the horizontal run, the arm of the framing square set at 12" gives us the level cut.

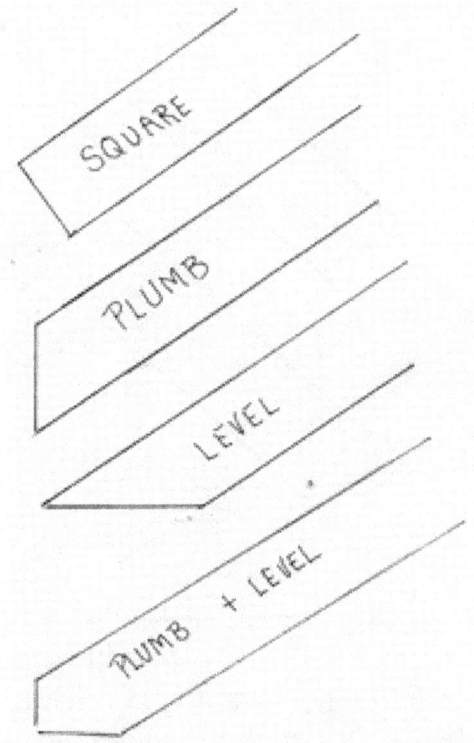

Different ways of cutting rafter tails where rafters overhang

If there is an overhang the rafter tails are generally cut either square, level or plumb. Knowing the pitch one can easily lay out the rafter tail or any combination of these.

If the rafters are either simply being butted against a ridge board or one another, a plumb cut at the peak is all that is needed.

In timber framing, more often than not, the rafters at the peak are likely to be lapped or bridled (*open mortised and tenoned*) to one another. This changes the angle of the rafter at the peak. It is neither a plumb nor a level cut and cannot be quickly drawn out using the framing square, unless of course the roof is a 12:12 pitch.

I generally use one of three methods to calculate the angle, depending upon whether I have the tools to calculate, the space to scribe fit, or the material to create a template.

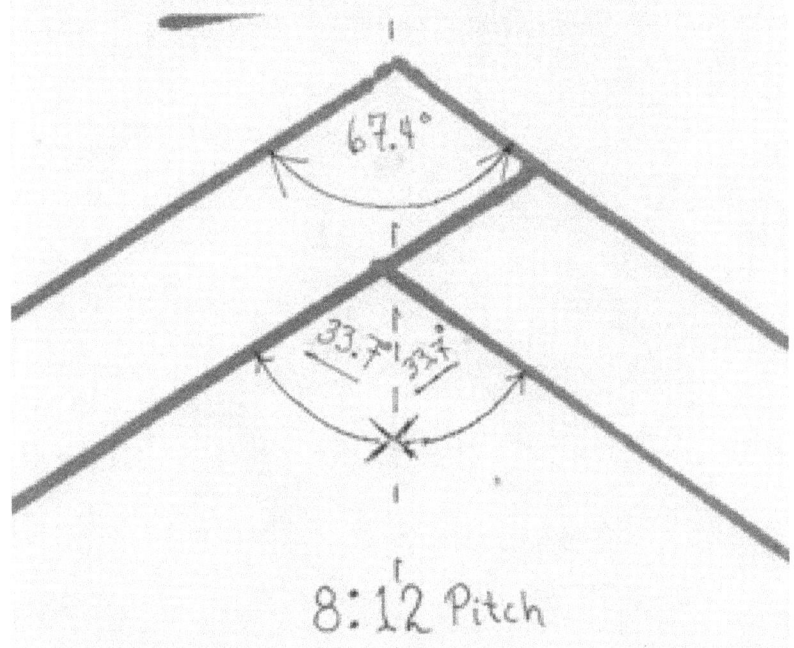

Calculating peak angles on lapped or bridled 8:12 pitch rafters

The ideal method is to use a contractor's calculator. If the pitch is entered into a contractor's calculator, the calculator will give the number of degrees from plumb the rafter is. In the case of an 8:12 pitch that angle is 33.7^0. Since each rafter of the pair is 33.7° from plumb. The angle of the rafter peak if bridled or lapped is 33.7° times two, or 67.4°.

Roofing squares, often called speed squares, have angle readouts and a pivot point that allows the user to draw a straight line at any angle.

8:12 Pitch, half-lap common rafters

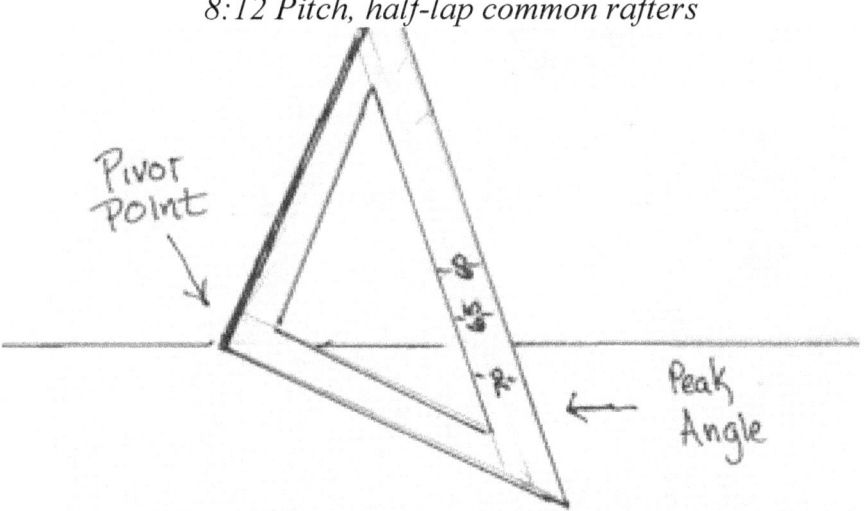

Roofing square or Speed square
Pivots at corner. Diagonal side shows how many degrees from zero the square is pivoted. When the square is pivoted to 67.4° from zero, the rafter angle can be drawn. With calculator and speed square the process takes but minutes.

Open mortise 8" pitch rafter with relief for small ridge timber

A bridled rafter pair: One rafter is tenoned, the other is an open mortise. This form of joinery is also called tongue and fork. A

skilled chainsaw operator can cut the open mortise entirely with the chainsaw. Or one can use a self-feed bit and bore out the back of the mortise then use a large circular saw and cut along the mortise lines. The central piece, after boring and sawing, tends to just fall out. A little clean up work with a chisel is in order whether one works with the chainsaw or the bit and saw.

I have used chalklines on the subfloor deck to layout the position of rafter pairs. I have laid the rafters along these lines and scribed the overlap of one rafter onto the other.

In contemporary light framing, the common way of joining the rafter to the plate is the *External Birdsmouth.* The external birdsmouth has been and continues to be used in timber framing.

The birdsmouth, once the pitch and rafter length are determined, can be laid out easily with the framing square. The external birdsmouth continues to be popular, because it is easy to cut, utilizes a common rafter system and easily allows for an overhang that can accommodate both soffit and a fascia trim. The biggest drawback to the external birdsmouth is that the joint itself does not resist outward thrust. The peg, spike or bolt used is the vehicle for resisting thrust.

External birdsmouth rafters being attached with "Timber Locks"

Layout for External Birdsmouth Rafter:

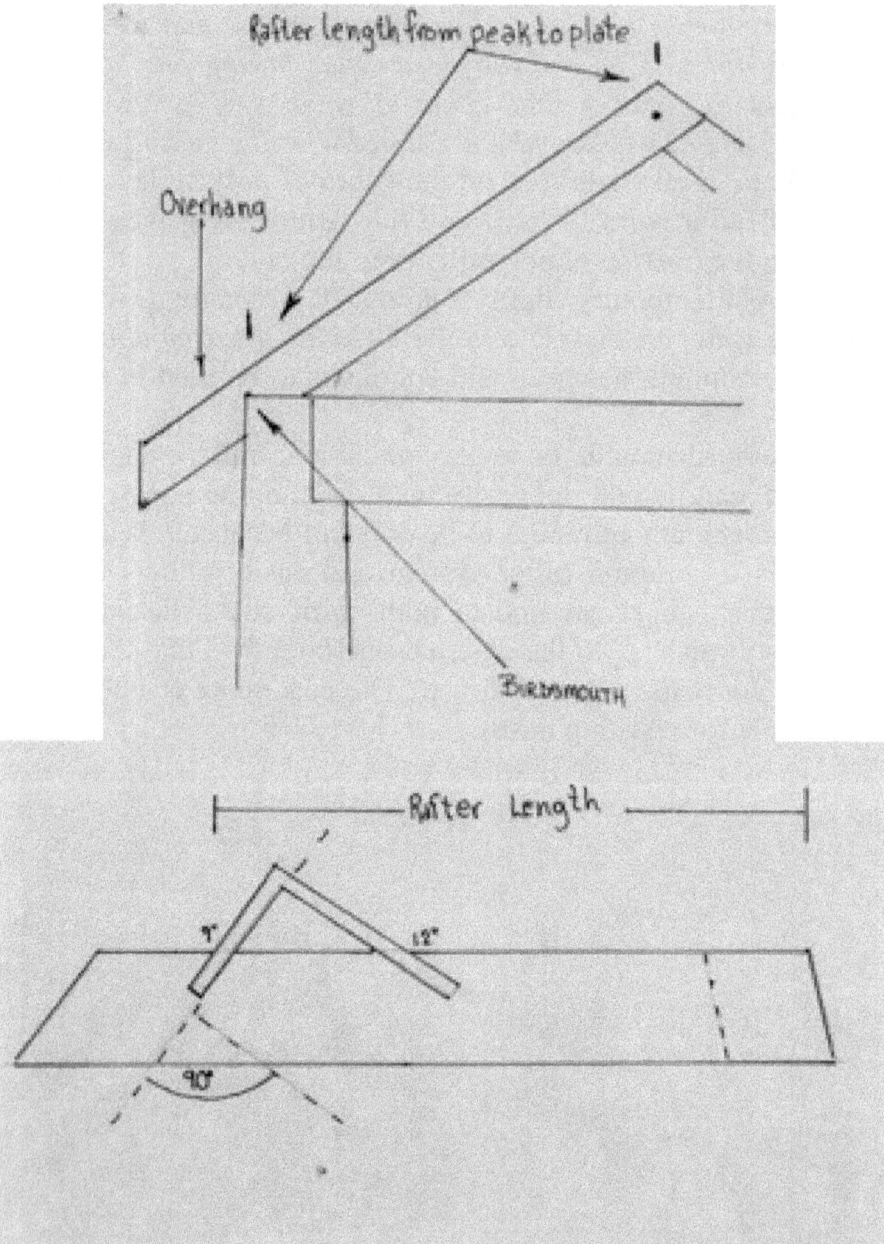

Laying out 9:12 pitch external birdsmouth. Plumb line is drawn using framing square set at correct pitch. Perpendicular to the plumb line a line is drawn outlining birdsmouth to be removed.

In conventional carpentry, the shelf created by the birdsmouth is equal to the width of the plate that it is to sit upon.

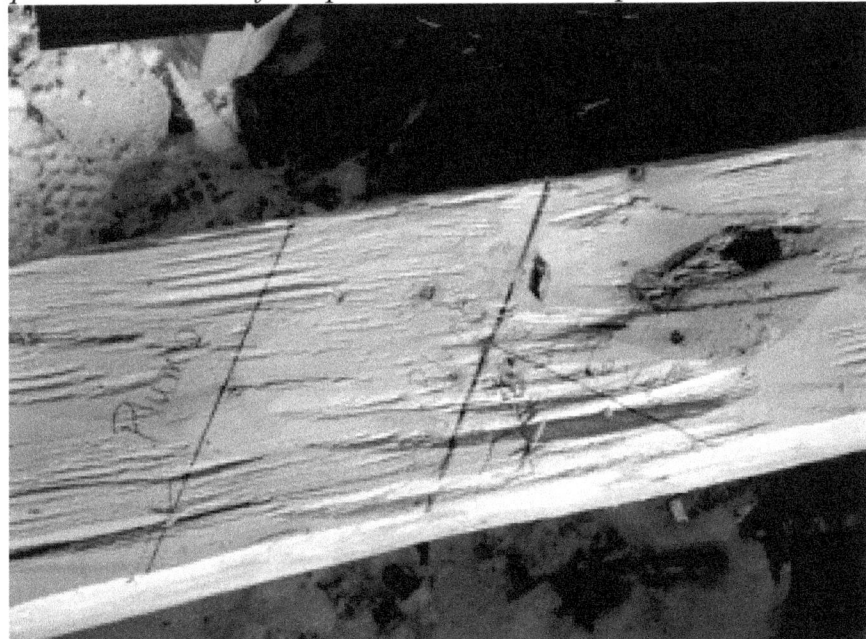
Birdsmouth laid out on hewn rafter

Finished birdsmouth

A stronger timber frame joint is the internal birdsmouth.

Internal birdsmouth fitted to the plate. The internal birdsmouth prevents the outward slide of the rafter.

Whether cut with a circular saw or chainsaw, either birdsmouth cut will have to be finished with a handsaw. Frequently the internal birdsmouth is used in a situation in which no overhang is to be employed, often where Structural Insulated Panels are used. The overhang is created by extending the panels not the rafters.

Laying out Internal Birdsmouth Rafter:

Different from the external birdsmouth, after measuring out the length of the rafter from the peak, align the square to create the level line rather than the plumb. Measure along this line equal to the width of the plate and draw a line perpendicular to the level line. Cut through the rafter along the lines to the point where both lines meet and the birdsmouth is completed.

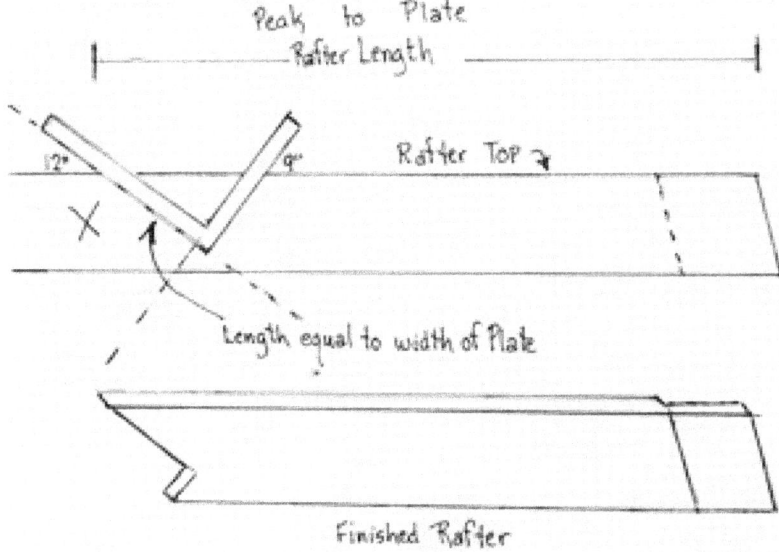

Layout for Internal birdsmouth

If one does want the common rafters to overhang but one does not want to use the external birdsmouth because of its inherent inability to resist outward thrust, the ***step lap*** rafter is a good option.

A small roof being built with step lap rafters

The step lap rafter also has it's own variations. One variation has vertical (or plumb) abutments, the other has square abutments.[7]

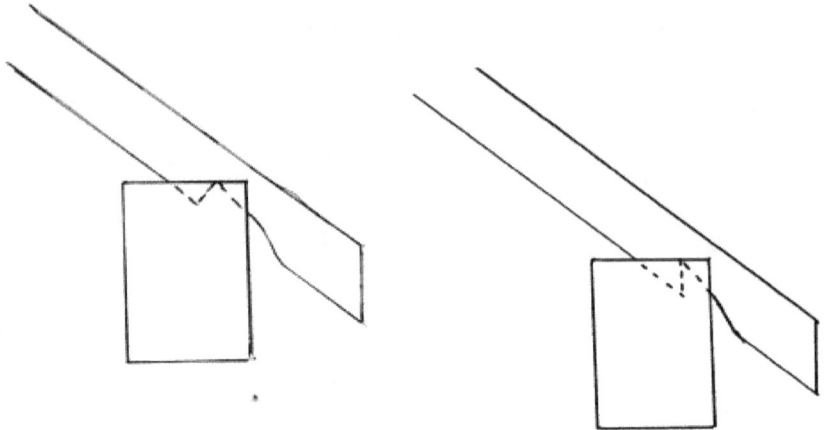

Step lap with square abutment.   Step lap with plumb abutment.

The photos that follow are of cutting a step lap rafter with a plumb abutment. Because, with a step lap rafter, the plumb cut is before the edge of the plate, when calculating rafter length to the step, the pitch has not changed but the run is shorter. If the step cut is 2" in from the edge of the plate, in a 9:12 pitch roof twenty feet wide, taking 2" off the run, reduces the rafter length to the step 2 ½".

A sliding bevel is very useful when cutting the mortise for the step rafters.

The first thing to do is set the sliding bevel to the proper pitch. The step layout is simple. It is simply two lines defining the width of the rafter, and a line perpendicular to these that defines the step. (in this case 2" from the outside edge of the plate).

The step can only be worked with chisels. One should outline the step with the chisel and mallet, Once the outline is cut to sufficient depth, the step is cleaned out and checked to see that the slope is correct with the sliding bevel.

---

[7] Sobon, Jack, *Historic American Timber Joinery: A Graphic Guide,* Timber Framers Guild: Becket, MA, 2002, pp. 34-37.

With a 9:12 pitch, a run of 2" from the edge of the plate means the sloping drop off the plate will be 1 ½".

Left—Cutting sloping plumb abutment with chisel.
Right—Checking abutment slope with sliding bevel.

Left—Outward slope cut with handsaw, sliding bevel acts as guide. It is cleaned out with chisel.
Right—Rafter tail plumb cut is done with a saw.

The sloping cut is easily cleaned out with an axe.

Rafter test fitted

Step rafters set into place on my brother Den's camp.

This rafter foot has both a step with plumb abutment and an internal birdsmouth. Accompanying plate below.

Birdsmouth and step rafters make sense when the rafters are joined to the plate. Frequently, however, a principal rafter system is used (English Tying joint, Truss systems, etc.) in which the rafters are joined to the tie beams, not the plates.

In such a system, there are basically two general methods of joining the rafters to the tie beams. One is simply to tenon the rafter and mortise the tie beam. The other is simply to build a matching beveled stop into both the rafter and the tie beam. Once pegged, this is a very effective means of securing rafter to tie beam.

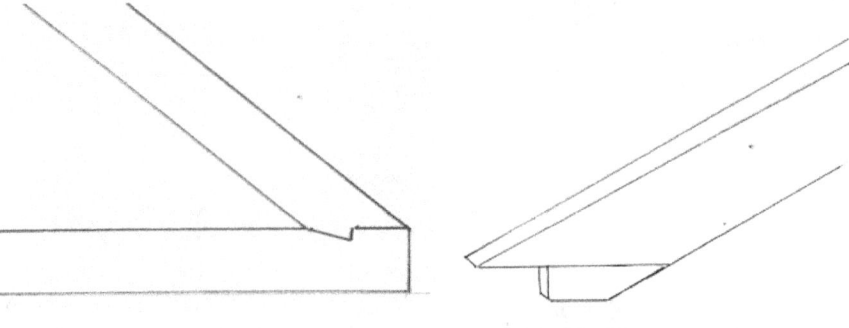

Rafter foot with beveled stop.   *Rafter foot with tenon.*

When the tie beam is cut to match the roof pitch, this length must be subtracted when cutting the rafter.
For example: if the roof length is 9'6" and the sloped cut of the tie beam is 9" than the rafter length is actually 8'9".

183

Aaron Sturgis has noted that in Southeastern New Hampshire there are two hundred year old structures still standing in which no joinery was used to hold the rafter to the tie beam. The rafters were simply level cut and joined to the tie beams with 1 ½" oak trunnels. The shear strength of a 1 ½" oak trunnel probably exceeds the shear strength of a rafter tenon.

If one wants overhang with a rafter / tie beam system, the easiest way to create it is to extend the tie beam.

Extending tie beam creates overhang

Rafter to post connection. This has become common in contemporary timber framing.

A rafter to post connection used by Maine Dept. of Inland Fisheries and Wildlife. Underside of rafter is mortised to fit post tenon. Post is set into rafter one inch.

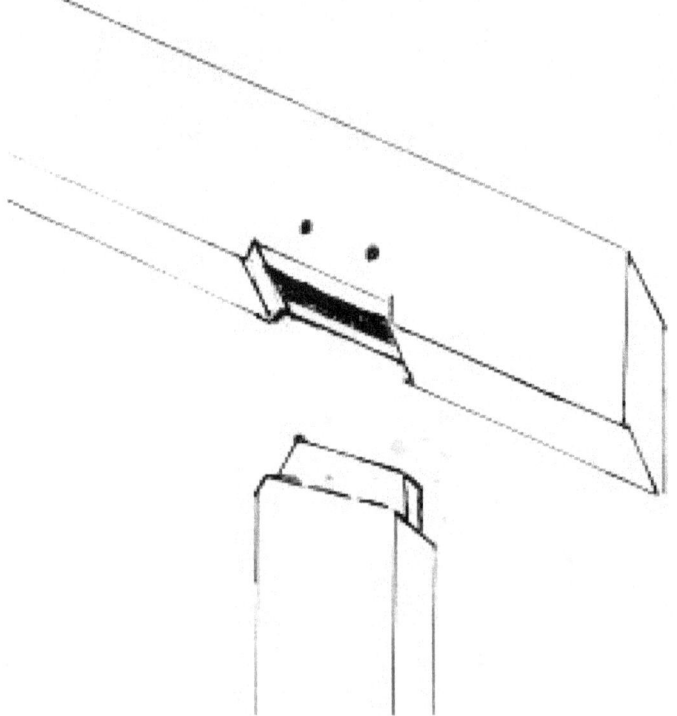

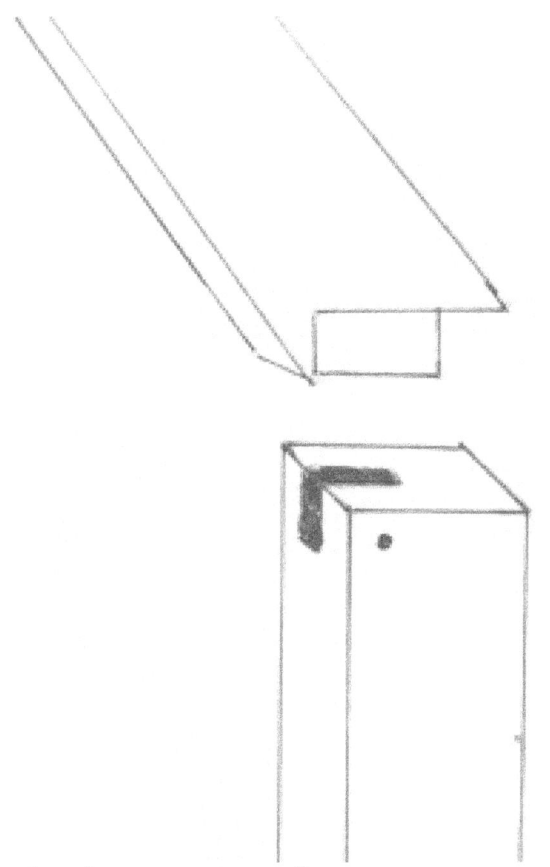

Another common rafter to post connection

With many different ways to attach rafters to a frame, it is not surprising that there are also many ways to frame a roof. In addition to the common rafter system, the principal rafter with purlins is perhaps the most common here in New England. Many of its early houses and barns were built using purlin roof systems.

In old Maine barns and houses, the top plane of the rafter was relieved to allow the purlin to pass through. This relief was often two inches deep. Today purlins are usually treated like floor joists. They either drop into square pockets or they are fitted to the rafters with dovetail joinery. Using dovetail joinery for the purlins creates a very strong roof system. I generally place purlins on 32" or 48" centers.

Principal rafter—purlin roof system

Purlins in this structure were left in the round with the top surface flattened.

Rafter laid out and cut with 2" deep pass through for purlins (Pownalborrough Restoration)

Full depth dovetailed purlin

Rafters laid out and cut with drop in purlins (Den Beaudry)

Principal Purlin mortised into Principal Rafters provides support for secondary rafters which in this case are pole rafters. Structure is a reproduction of 17th century dwelling cut and raised by Holder Bros., Monroe, Georgia

Another common system, used frequently on Yankee style barns from the mid 1800s on, is a Principal Rafter—Principal Purlin roof system with secondary (or common) rafters. Both of these systems are also used with *Truss* systems. Two common Truss systems are the King Post and The Queen Post Trusses. I find the King Post Truss very compatible with the Purlin roof system.

Tie Beams are limited in their ability to span long distances. Generally an internal support post or two must be placed under the tie beam to keep it from sagging, or one can use a Truss system. The truss system works in tension and uses the stresses created by the roof to keep the tie beam from sagging without internal supports.

The simplest truss: a collar tie added to rafter pair.

King post truss

190

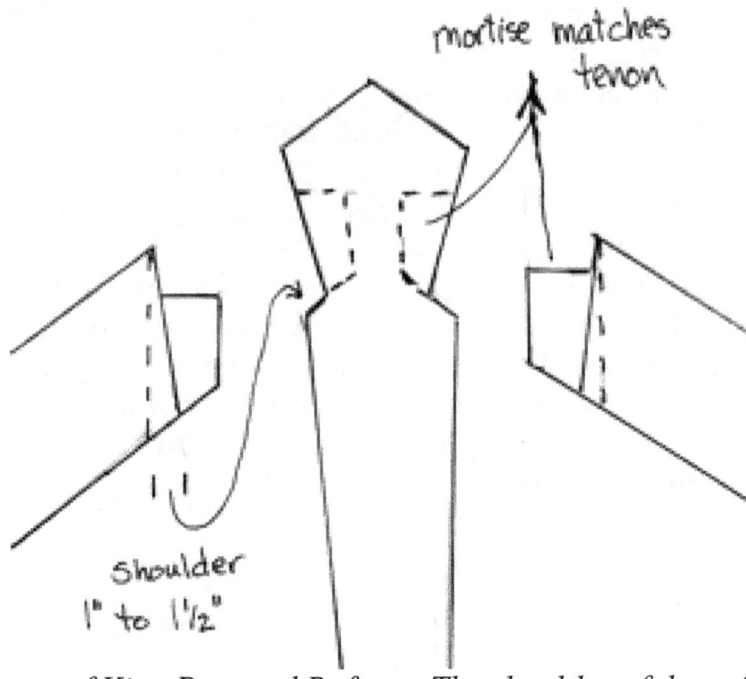

Diagram of King Post and Rafters. The shoulder of the rafters is 1" to 1 ½" off plumb. Tenon is at least 3" long to accept trunnel. Top of King post matches roof pitch and forms part of diagonal run when calculating rafter length.

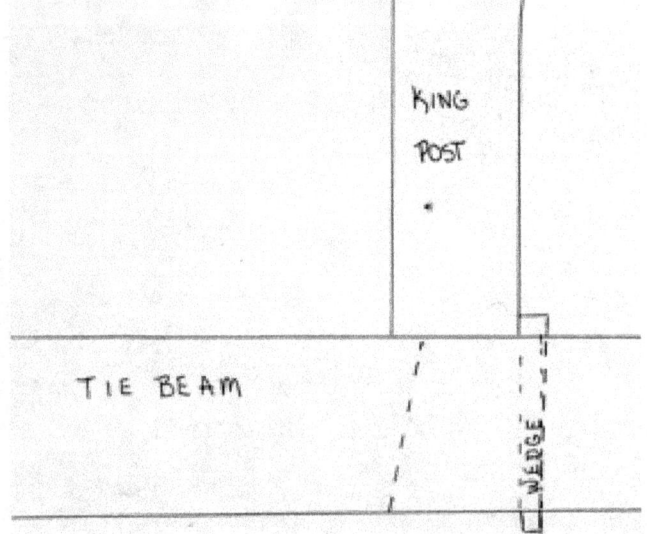

King post is dovetailed, wedged and pegged to tie beam. (See instructions for Dovetailed tie beam in "Tying it all together.")

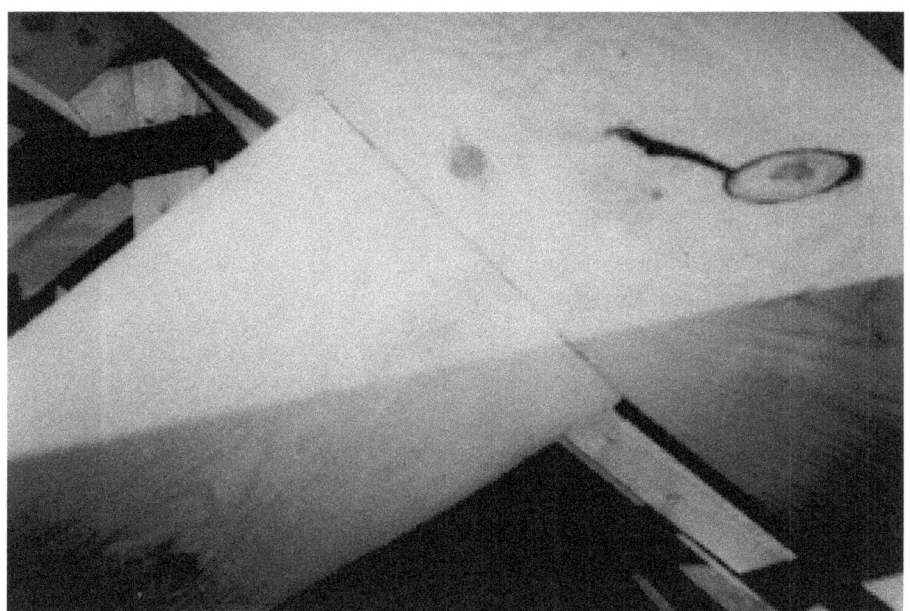
Truss strut is tenoned, pegged and shouldered.

With a King Post Truss, the outward thrust of the rafters is transferred along the tie beam to the King Post and Struts. As the rafters push outward, the king post is pushed upward taking the tie beam with it, keeping the tie beam from sagging. Clear spans of forty feet have been accomplished with the King Post truss and even longer spans with the Queen Post truss.

The base of the king post is not simply tenoned to the tie beam. It is dovetailed, wedged and trunneled. Being in tension, the king post does not merely rest upon the tie beam, it is actually holding the tie beam up. The rafters rest into the king post with shouldered tenons. The struts are usually shouldered and/or tenoned. If the frame has a small ridge timber, the king post is notched to receive it.

The Queen Post Truss and a modification of it called the Queen Rod Truss appears to have been very common in the late 1800s, which is not to say that it was not in common use before this time. My experience with them is with barns and carriage houses of the Victorian age. On one barn I was asked to repair the trusses and completely rebuild two. I looked at the trusses with Rick Irons, the general contractor. He asked me if I had ever repaired a queen post truss. I said no, but reassured him that it wouldn't be a problem because I had done king post trusses and this queen post truss was simply a king post truss cut in half and stretched with a horizontal member.

Queen Post Truss.
Principal rafters of truss did not go all the way to peak, but rather formed the support for the principal purlin that was to carry the secondary rafters.

This Queen post truss is very much like half a King post truss.

Queen post truss and Principal purlin frame.

The Queen Rod truss simply substitutes metal rods that are secured in tension to the frame in place of queen posts.

The roof system on the above barn employed both Queen post and Queen rod trusses. Principal purlins tied the trusses together as well as supported the common rafter system.

One ancient, medieval roofing system that is making a comeback is the cruck. This system is given a chapter all its own.

Cruck Framing

Cruck framing is an early Medieval form of framing dating back in England to at least the 12th century. *Cruck* is simply a word meaning crooked. It refers to the fact that the cruck pair was derived from a crooked tree. By the time the colonists had arrived in North America, its use had all but vanished and it was not brought here as a building style.

The recent revival in timber framing has created a new interest in this archaic method of framing. Its long sweeping forms give a softer organic flare to a building system that can appear all too linear and rigid.

The cruck is really a roof system within a roof system. It is a curved A-frame with a second set of rafters resting upon it.

The strength of the cruck is in the fact that the downward thrust of the rafters is transferred along the crucks directly to the foundation rather than along the rafters to the plates as outward thrust. The cruck by its very nature eliminates the need for a tie beam, requiring only two small cruck spurs attaching the outside post to the crucks.

The cruck by its angled nature also eliminates the need for diagonal bracing along the gable walls. Having neither tie beam nor braces gave sufficient space in the gable of this log building for an arched window sitting much higher than would have been possible had a tie beam been used.

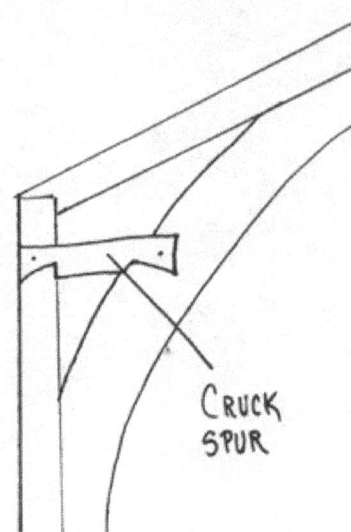

Cruck frame above hewn log addition. Post are tied to cruck by half-dovetailed hardwood cruck spurs. The spur eliminates need for tie beam.

Even though a cruck system requires no tie beams, one may still choose to have tie beams. In Robert's cruck frame house, every cruck pair is joined to a tie beam.

Each cruck is an individual expression. One cannot simply go to a lumberyard or sawmill and order one. The process starts in the woods. One must find a tree with the proper curve and enough mass to yield a matching pair.

Robert found on his property some large arching oaks growing on the pasture edge of his woodland. Two of these oaks had diameters of around 15 inches. Moe Martin, a woodsman and sawyer, felled the trees and with tractor and loader, moved them to his portable sawmill. It was all his loader could do to move these 20' long oak logs.

Even after sawing, the six-inch thick oak crucks were far too heavy to be handled without equipment. The cruck pairs were brought over to the site and stickered near the main deck.

Lines were snapped on the subfloor representing the interior outline of the post, tie beams and rafters.

Oak log positioned upon sawmill for sawing out cruck pair.

Too heavy to lift, pry bars and rollers were used to move the crucks into position for scribing and cutting.

Cuts were made with the 16" circular saw. The first was at the base of the cruck, cutting along the line that forms the intersection of post and cruck. The second and third cuts were at the peak and form the intersection of the rafters and cruck.

Once these cuts were completed, the second cruck was maneuvered over, lifted unto cribbing, aligned atop the first cruck and leveled.

With the crucks lapped and fitted, the lines that form the intersection of the floor and the cruck were scribed onto the base of the cruck. A tenon was laid out directly beneath this line and the tenon was shaped The pair was then disassembled and set aside as the next pair was worked.

When all three cruck pairs had been scribed, cut and set aside, I went over my chalklines on the subfloor with pencil to make them more durable. The crucks would have to be laid out again to fit tie beams and collar ties. In addition to the above lines, a centerline running from the point of the peak to the midpoint of the post was snapped.

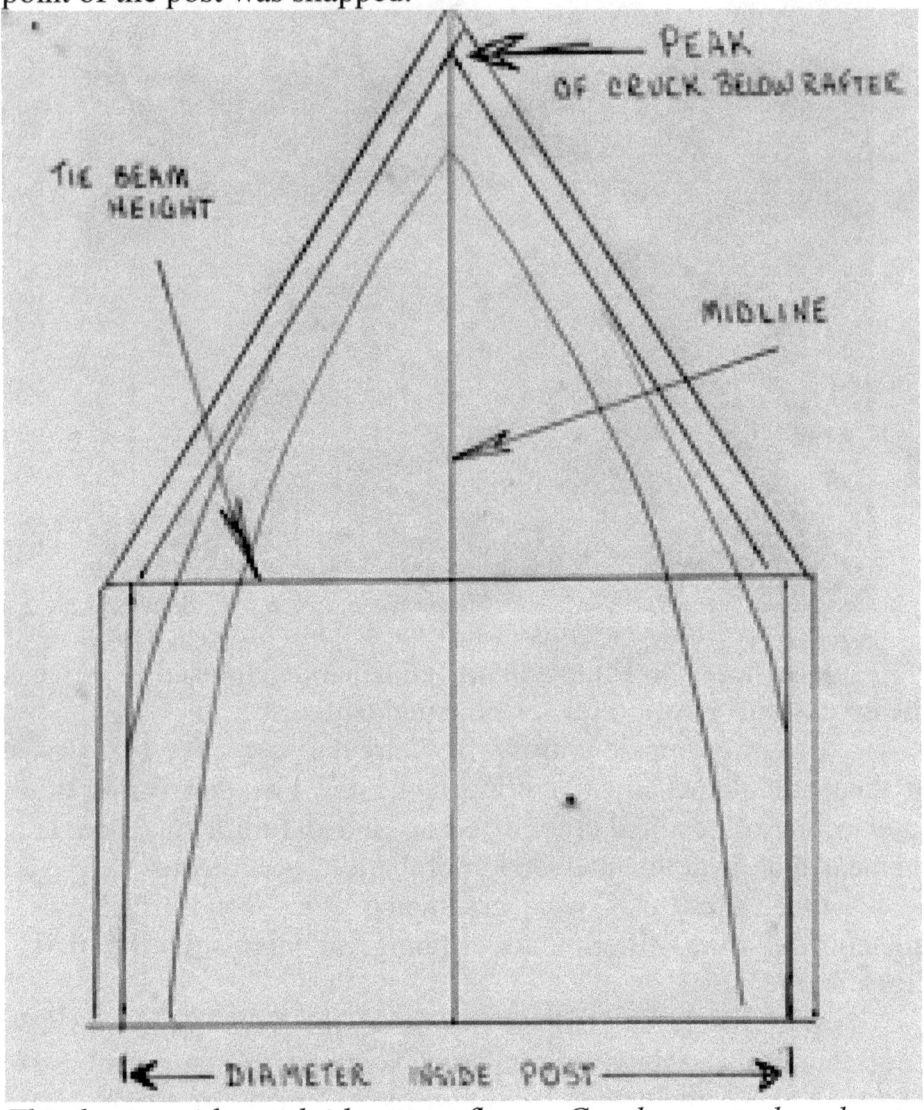

This basic grid was laid out on floor. Crucks were placed out along the outline. The lines established intersections to be cut. Grid also served as layout guide for joining tie beam to cruck.

1st cut – (to left of saw) defines intersection with wall post
2nd and 3rd cuts define intersections with primary rafters

Top cruck in position for scribing,

Cruck pair now ready to be matched together with Lap.

Cruck pair lapped and fitted

The remaining bark was peeled off the crucks. The crucks were planed, sanded and oiled, along with the tie beams and collar ties. Again, using level and plumb bob, a cruck pair was positioned and plumbed to the lines on floor. Peak was joined and the crucks were leveled on cribbing. Collar ties and Tie beams were placed, scribed, cut and fitted to the crucks.

Collar tie in position to be scribed on the top cruck pair. The board across cruck pair in front of collar tie, was used to check that cruck pair was sitting level. The collar tie on the bottom pair is fitted and trunneled.

Tie beam being cut to match reliefs cut into cruck pair. Lower cruck shows a fitted tie beam.

Finished cruck pair with tie beam and principal rafters attached. The placement of tie beam on cruck pair is critical. The midpoint of the tie beam is aligned so that it intersects the centerline running from peak to base. The ends of the tie beam have to meet the plate so that once fitted to the cruck the cog on the tie beam sets into the mortise on the plate.

When set between plates, the cogs engage plate mortises.

Tie beam intersects cruck, rafter and plate.

Crucks carry the weight of the frame and provide strength, in particular, to places where the frame itself might fail. Sometimes, however, a cruck serves no other purpose than beauty.

Scarf Joints

The distances that must be spanned frequently exceed the lengths of available timbers. One solution is to scarf two timbers together. Scarf joints vary greatly. Some are quite simple, particularly where the scarf is directly supported by a post. Others, however, are complex, but are almost as strong as a continuous beam. Scarves such as these are self-supporting, requiring no post beneath them.

I enjoy varying the scarf joints I use, mostly because I enjoy cutting them. Perhaps the simplest scarf joint I have ever used was a simple half-dovetail lap. I used this scarf in a log home in which the logs were uniform and flat, top and bottom.

A simple half-dovetail lap used in log construction.

The scarves of each course of logs were staggered by several feet from the scarves of the preceding course. At the scarf, the logs were feathered with plane or chisel to give a near seamless appearance to the log wall.

A simple scarf is the *Stop-splayed* scarf. This is a very common scarf.

Stop-splayed scarf

Generally where I have seen this scarf used, it has been supported with a post beneath it. I tend to use it on sills where the scarf is supported by the foundation wall.

The overall formula that I generally use, though not always, for timber frame scarves is that the scarf should be 3 times the thickness of the timber.

The formula for laying out the stop-splayed scarf is not complicated. Assuming the timber being scarfed is an 8" x 8", the scarf will be 24" long. **Scarf length is 3x the diameter of timber.** Because sawn timbers can vary in width and thickness, it is important that the measurements be made from a reference face. In the case of a sill, the reference face is the bottom.

Steps for laying out stop-splayed scarf.
Step 1: Using the square draw a line 2" up from bottom onto timber end.
Step 2: Measure over 24".
Step 3: From this point measure up 6". Draw a line connecting this point with the point in Step one. This creates the splay.
Step 4: Finish the laying out of top stop with framing square.
Step 5: Repeat the process on the opposing side.

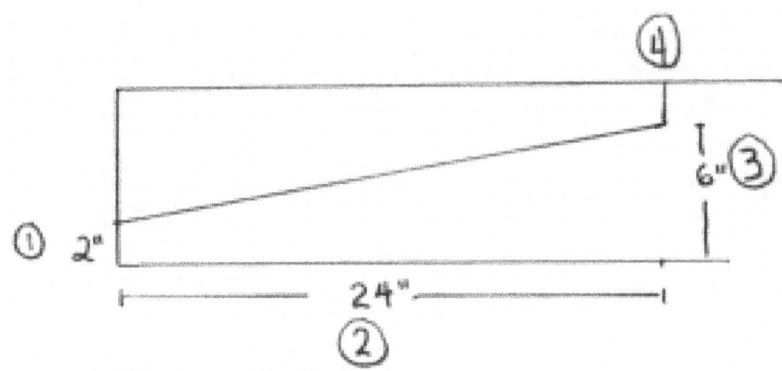

Diagram of steps 1 thru 4

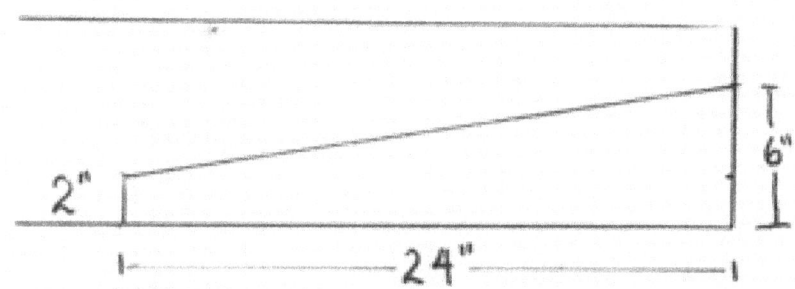

Adjacent timber is a mirror image.

The scarf can be cut with a handsaw, a chainsaw, or a circular saw. If the circular saw is used, because of its depth limitation, one will have to cut along the splay line, rotate the timber 180 degrees and cut along the splay line again. If the saw is a large 16" beam cutter, the two cuts will meet and only a small portion at the end will have to be cut with the hand saw. If the saw is a 7 ¼" saw, there will still be 3" of wood between the two saw cuts that will have to be cut with a handsaw or a reciprocating saw with a long blade.

If the chainsaw is used, it is best to stay slightly away from the line and clean it up to the line with the electric plane, chisel or slick.

The final step is to use the framing square to check that the scarf has no high points that will impede it properly seating. Ideally, the framing square should only touch the timber on its outside edges. It is desirable to have the surface slightly dip in the middle. When wood seasons, the outside of the timber shrinks more than the center. If the scarf is perfectly flat, as the outside edges shrink, the center will start to bow up. If a slight concave ellipse is created, even as the outside shrinks, the center will never rise above the plane of the scarf. This applies to all scarves.

Placement of trunnels in stop-splayed scarf

When joined the scarf is held with four trunnels. Trunnels are placed 2" from the edges of the timbers, 4" from the stops.

Many timber framers come to rely upon a particular scarf. The reasons can vary. One may simply enjoy cutting it, or enjoy the appearance, or it may simply be the best at meeting ones needs. It is, however, good practice to be familiar with more than one scarf.

There is one that I particularly like. It is the *Splayed-scissor* scarf. I came across this scarf in 1978 in *The Timber Framing Book* by Stewart Elliott and Eugene Wallas. I have never seen this scarf used in any old structures around the area in which I live, but it appeared as a scarf that would be very strong at supporting itself. The layout is simple. It is best cut with a large circular saw. It is simply two identical right triangles that meet at a centerline. One slopes up, the other slopes down.

Scissor Scarf

The steps for cutting are equally simple. I set the 16" circular saw at full depth and cut along the midline for the whole 24". The log is turned 180 degrees and the cut repeated.

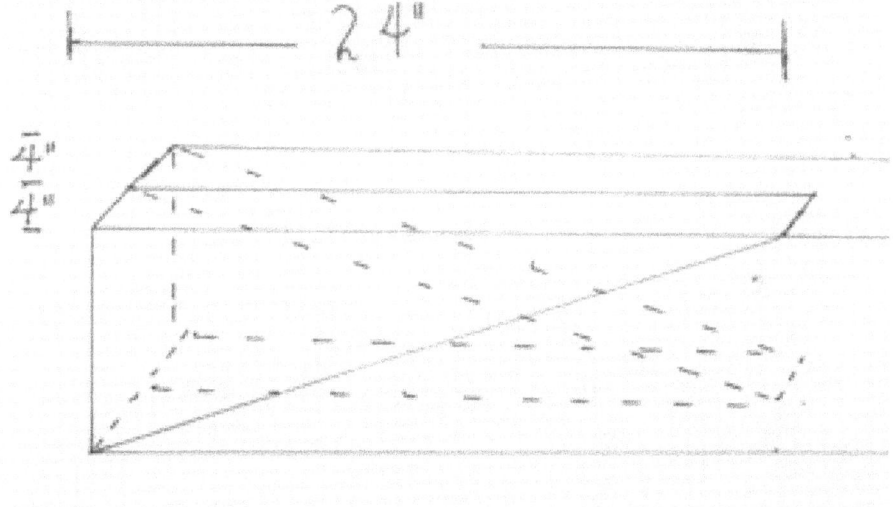

Cross section diagram of scissor scarf

I then rotate the log 90 degrees. This places the broad surface of one of the scissors facing up. I set the saw depth so that the saw cuts only to the midline cut. I cut along the splay line and one triangle falls off.

I rotate the log 180 degrees. I again cut along the splay line, removing the second triangle. What is left is the scissor-splayed scarf. The timber to which it is to be joined must splay in opposite directions. The scarf, frequently, must be pulled together with come-a-longs and straps. I use four trunnels to hold it.

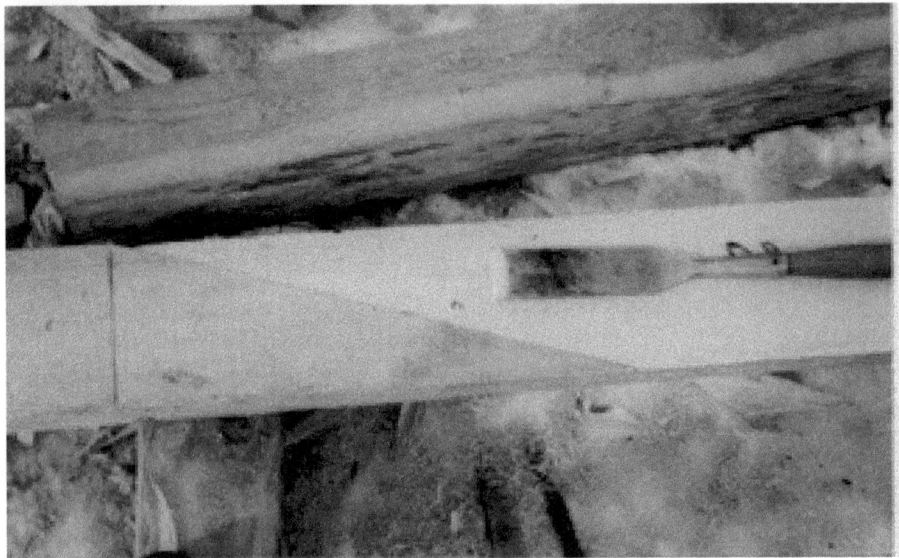

Fitted Scissor scarf

Scissor scarf used in log construction

This scarf, however, is not useful in all situations. Whereas some scarf joints are designed so that one timber drops into another, the scissor scarf requires that the timbers must be slid in to one another. For this reason, in many applications, it must be joined prior to raising it into place.

If one needs a scarf in which one section can be raised onto the frame and the other section dropped in from above, a good scarf is the *Halved and Bridled* scarf.

Essentially the Halved and Bridled scarf is a long half lap with both a tenon and a mortise. The layout and cutting is done in two steps. The first is to cut the elongated half lap. To allow for the tenon that will be fashioned, I cut the lap 28" long. I create a centered two by four inch tenon on the tongue of the lap and bore and chisel a centered two inch by four inch mortise into the timber immediately beyond the lap. Its complement is made on the other timber. When fitted together, its side profile will show only the 24" half lap.

The scarf is held together with 6 trunnels. Four go through the lap. Each should be 4" from the stop, 2" from the edge of the timber. One each is also placed through the bridled mortise and tenon on each end of the scarf.

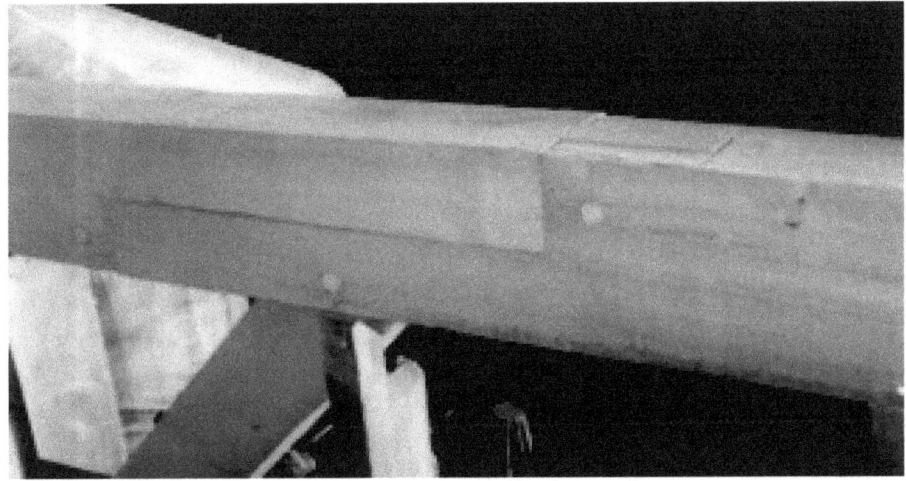

Halved and Bridled scarf

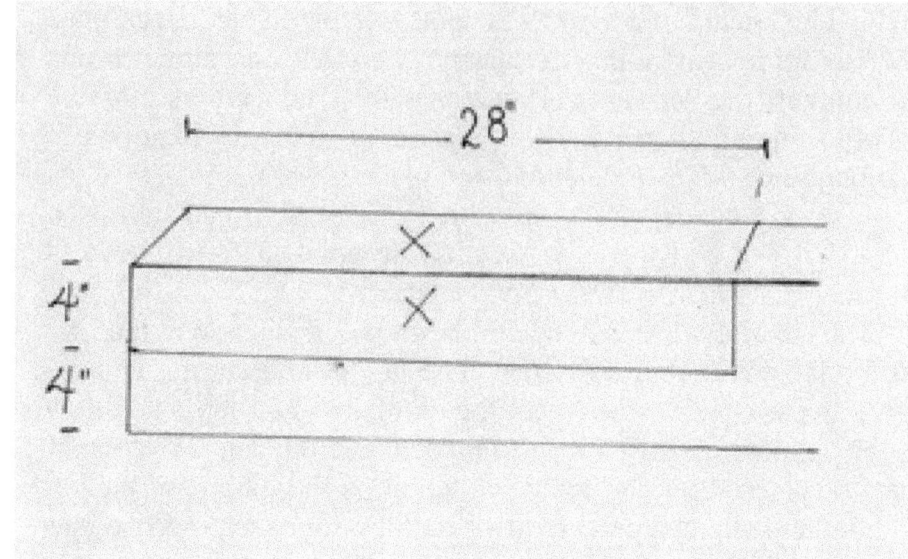
Step 1: create a 28" lap.

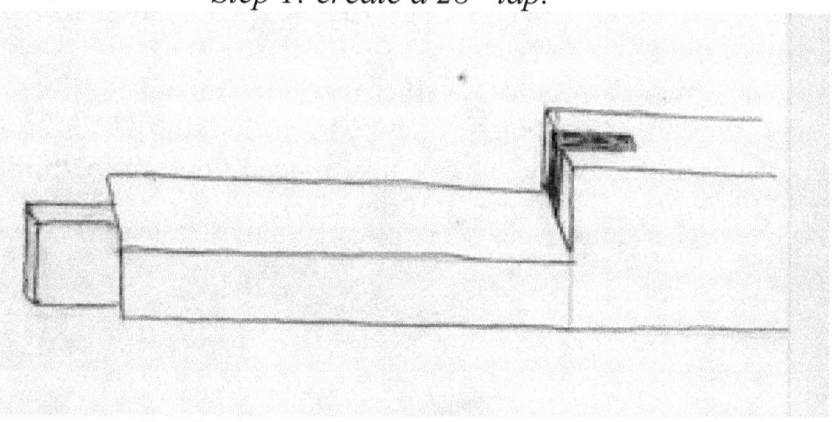
Cut both a 4" tenon and 4" mortise.

An invaluable scarf I learned from Aaron Sturgis of Preservation Timber Frames is the *Bladed* scarf. Not only is this an effective scarf for bridging long spans, it is an excellent scarf for post and timber repair.

Unlike the bridled scarf, the bladed scarf simply has small 2" long stub tenons that run parallel to the lap rather than perpendicular to it. The overall length of the scarf, including the stub tenons is 28". On the tip of the lap one must fashion the stub tenon and at the lap ends one must fashion the mortise for the stub tenon. The stub tenon, though only 2" long and 2" thick, runs the entire width of the timber. As with the other scarves, I peg with 4 trunnels, each four inches from the stub tenons, two inches from the outside edges of the timbers.

Bladed scarf is an excellent scarf for timber repair

The color contrast between these two timbers makes the layout of the bladed scarf readily apparent.

Cutting a Bladed Scarf:

Step 1: Rip along mid line for half-lap of Bladed scarf. (Small rip-cut to the right is for the tenon or blade.)

Step 2: Outline mortise with framing chisel before boring.

Step 3: Bore out mortise

Step 4: Cut off the half abutting mortise, creating lap.

Step 5: Clean out mortise, shape blade. Check that lap has no high points.

There is one more scarf that I use. It is called the *Stop-Splayed, Undersquinted and Tabled with Wedges and Pins* scarf. I will simply call it the Tabled scarf.

By tabled one simply means that the splay is divided so that one half sits deeper than the other half. By squinted one means that the stops are not perpendicular to the reference face of the timber. I usually cut the stops perpendicular to the splay. Once pulled together the squints resist slipping up or down in the scarf.

The deeper of the two tables is also longer. This creates a space to drive in two opposing wedges. The wedges drive the tables tight against the stops.

Stop-Splayed, Undersquinted and Tabled with Wedges scarf.

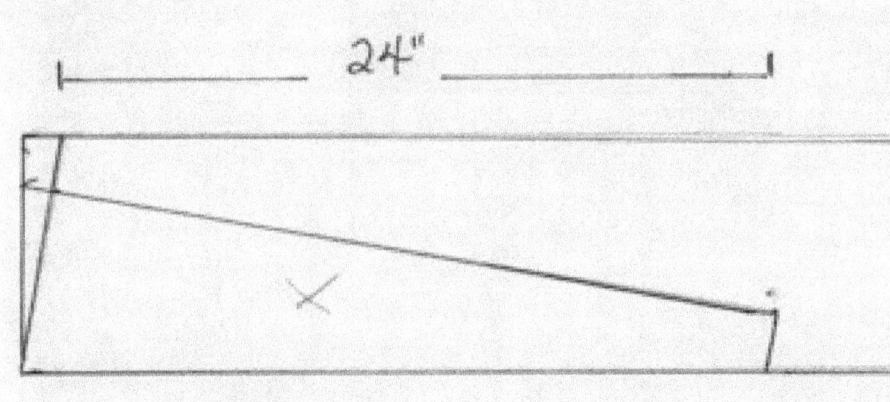

The first steps in layout and cutting of the tabled scarf are the same as the layout for the Stop-splayed scarf, only the stops are squinted. (I generally cut the stops perpendicular to the splay.) On the Tabled scarves I have seen, the stops are either 2" or 1 ½" in depth. Cut the splay and stops and smooth the splayed surface. Once the surface is clean and straight the tabling of the splay can be laid out.

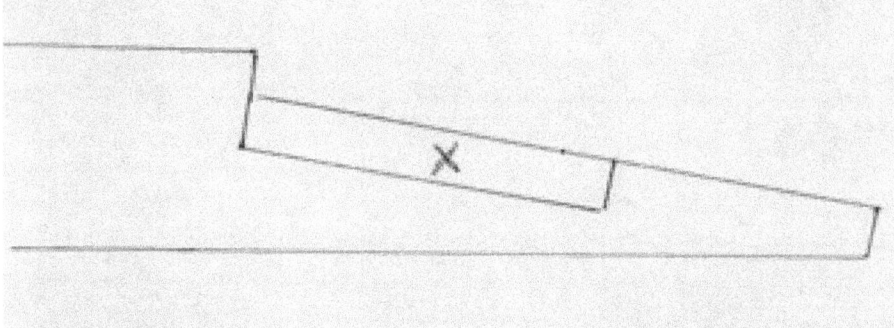

The second step in layout and cutting of the tabled scarf is creating the tabled surface. The tabled surface is half the length of the splay plus 1" for the wedges. (This creates a 2" space for wedges once the two timbers are joined). The depth the table is dropped is the same depth as the stops. If one used 2" for stops, the table drops 2".

If one has a chain-mortising machine, it is possible to bore the tabled area out. If not, the usual method is to cut *several kerfs with the circular saw and clean it out with the framing chisel. It is usually finished with the slick or a plane. Check again with the framing square for any high points.*

The wedges are simply two long triangles of hardwood. They are the same height as the stops. The two wedges are pushed in at the same time from opposite sides. They are then tapped deeper with a mallet until the stops are pressed tight against the ends of the scarf. A word of caution: Once test fitted with hardwood wedges, this scarf can be very difficult to disassemble.

In addition to the opposing wedges, the scarf is also held together with 4 trunnels: each 4" from the stops, 2" from the edges of the timber.

This scarf can also be done with square rather than squinted stops.

220

The scarved plate being set into place. The square stops allow the top half to be dropped onto the bottom half.

Splayed Tabled scarf on an old Norwegian log building. Open air museum, Lillehammer, Norway.

This splayed tabled scarf had neither wedges nor stops and clearly was placed near an internal wall partition. I asked about it and was told that the tabled scarf was a common scarf in Norwegian log construction. A clear reference is made in Praktisk Lafting by Hogsnes, Leine og Sageng.

Of course, despite all the scarves one knows, there always comes a time, especially in repair work, where none seem to fit. At times like this the only thing I can suggest is be creative. Take what you know of joinery and what you understand of the forces and stresses placed upon the timbers and create something.

Shingles and Trunnels

The first building I roofed with hand-split shakes was only eight foot square. It was built in the early 80s and the original shakes were still on it when I took the building down last year.

I had earlier visited a friend further up the road and he showed me how he had covered an out-building with crude shakes he had split from poplar. I commented that I believed poplar was not durable enough for shakes. He answered that poplar was what he had and he believed, being used on the walls rather than roof, the shakes will last a long time.

In the early 80s poverty was my common lot. A newly planted back-to-the-lander, I worked for small wages and learned to make do. I owned a froe, so I went out and made a froe club and commenced to learn shingle riving. I experimented with what I had – Eastern White Pine.

The amazing thing about splitting out shingles is how few tools are actually needed. The principal tools for splitting out shingles are the froe and froe club. I, also, find it handy to have a hewing hatchet on hand.

Left -- My first attempt at shake making. Pine shake roof built in the early 80s. Original shake roof was still on when building was taken down in 2008.
Right – Tools of the shake maker: froe, froe club, hatchet

My early shakes were fat, generally an inch or so thick. With time I have learned to create a more uniform, thinner shingle. After roofing the small outbuilding with shakes, I tackled covering the walls of the house with shakes, another outbuilding and a small barn.

Small barn roofed with pine shakes.

Many people question my use of pine shingles. My house walls have had hand-riven shingles upon them for around 25 years. As siding I believe they will easily last 40 years. I have not used hand-riven shakes to roof my house, but I do use them on my outbuildings. Without treating with oil, I get about twenty years of use. I do not know how long oil treated shingles will last, but I have since adopted the practice of treating with linseed oil. I do know, however, that in the 18th century, at least in Maine, pine shingles were taken in barter for other goods.

Once one gets the hang of it, pine shingles are fairly simple to make. It takes clear wood to make shingles. Getting clear shingle wood with pine is easy. Pine has a unique way of throwing out limbs. At the start of each growing season, the white pine throws out at its tip a whorl of branches. Since the branches always begin as a whorl at the tip of the tree, the branches of the pines are separated from one another by a year's growth. The area between the whorl of one year and the whorl of another year is free of limbs, therefore clear. The length of these clear sections vary from as little as twelve inches to as long as thirty six inches, depending upon the growing conditions the year that length was formed. I keep my shingle bolts to around 18". To do so, I simply fell a pine, then walk up the tree, cutting 18" sections from between the knots.

In addition to being knot free, the shingle bolt must be straight and cannot have spiral twist. Shingles split from wood that has spiral twist will all be twisted. Unlike sawing, with splitting the shingles follow the grain. If the wood grain has twist or bend, the shingles will likewise.

Actually, most of my shingle wood comes from harvesting pines for other reasons. I fell a pine because I need a timber. Sections of wood below and above the piece I intend to hew are cut out for shingles. The shingles are then rived, shaped and stored for when they will be needed.

Surprisingly small pine trees can be used in shingle making. Shingle bolts in the 8' to 12" in diameter work perfect for me. I have used bolts as small as 6" in diameter. Bolts larger than 12" are generally split and quartered first with maul and wedge.

The process itself is rather simple. A shingle bolt is stood up upon a block of wood. The froe blade is balanced across the midline of the bolt and it is driven in with the froe club.

Left -- Place blade across center and drive the blade in. Pull handle back and bolt should pop into two halves.
Right – On difficult splits use knee as brake and pull up.

227

*Away from the center pith the wood splits clean. Keep **halving the halves** until the desired thickness is achieved.*

The final split is all finesse. I use my knee, and work the froe gently forward and backward trying to avoid having the froe run off and ruin one of the shingles. With the hatchet I trim off the bark.

I also trim off any high bumps with the hatchet.
Of late, I have started shaving taper using a shavehorse.

Once the entire width of the blade has been driven in, a quick pull or push of the handle usually splits it. Wood around the center pith is generally brittle and may split in a ragged fashion. Ignore it for now.

It is a process of halving and halving the halves. Set aside one half and stand the other back up upon the block. As before, place the blade at the midline and drive the froe blade in. Once the blade is driven in, I often find, to increase leverage, I brace the wood against the block by pressing my knee upon it while pulling up on the froe handle.

The process of halving continues. When a section is approximately the thickness of two shingles, one does the last split. The froe must be carefully centered. At this point the froe blade is not so much pounded in as rather tapped in. This is the one point in the splitting process where to a certain extent we can actually influence the split. I put down the froe club, grasp the shingle with my free hand while attempting to guide the split by pushing and/or pulling the froe handle. One wants to prevent the split from running off, thereby ruining one of the shingles.

The central section containing the center pith is split down the pith and smaller shingles to the right and left of the pith are made.

For years I was content, once I had trimmed the edge with the hatchet and hewn away any high spots, to call these shingles finished . I have added a last step.

The shingle is placed into the jaw of the shave horse. The shave horse is a very ingenious, foot-operated vise for working with the drawknife. Once the shingle is held, the drawknife is used to shave, particularly the underside, to a smooth, flat, taper.

My usual method is to work with the froe and hatchet until several shingles have been split. These are brought to the shave horse and shaved. Then they are stacked to dry.

I have never made shingles for sale or barter. I make shingles for the structures upon my own property. Hand made shingles have a slightly wavy or irregular appearance. This gives a more natural, organic look to a building. Shingle making is a simple activity that requires very little mental concentration. One works with one's body. The physical labor can be relaxing. Putting aside an hour here or an hour there, one can quickly accumulate a sizeable number of shingles.

Trunnels are also easily made with froe and shavehorse. Trunnels are as necessary to timber framing as nails and adhesives are to conventional framing. They can be purchased or in a pinch dowels can be used. I, however, always make my own. Trunnels are generally made from straight grain hardwood. Since trunnels are usually from 12" to 18" in length, the easiest way to get trunnel material is to pull it from the firewood pile. Oak and Ash make good trunnels. I have made them from other woods such as locust, but oak and ash are readily available from my woodlot, and both make excellent trunnels.

The process for creating trunnels is slightly but not all that different from splitting shingles. I generally quarter the piece with a splitting maul. I place a quarter upon the block and using a combination square or 1" wide rule create a grid of 1" squares. The froe is placed upon one of the pencil lines, preferably near center and the froe driven in. I continue picking up pieces and splitting along pencil lines until each piece is roughly 1" square. Not all splits will run true and some pieces will have to be discarded. An 18" diameter round of wood could yield as many as 70 to 80 trunnels.

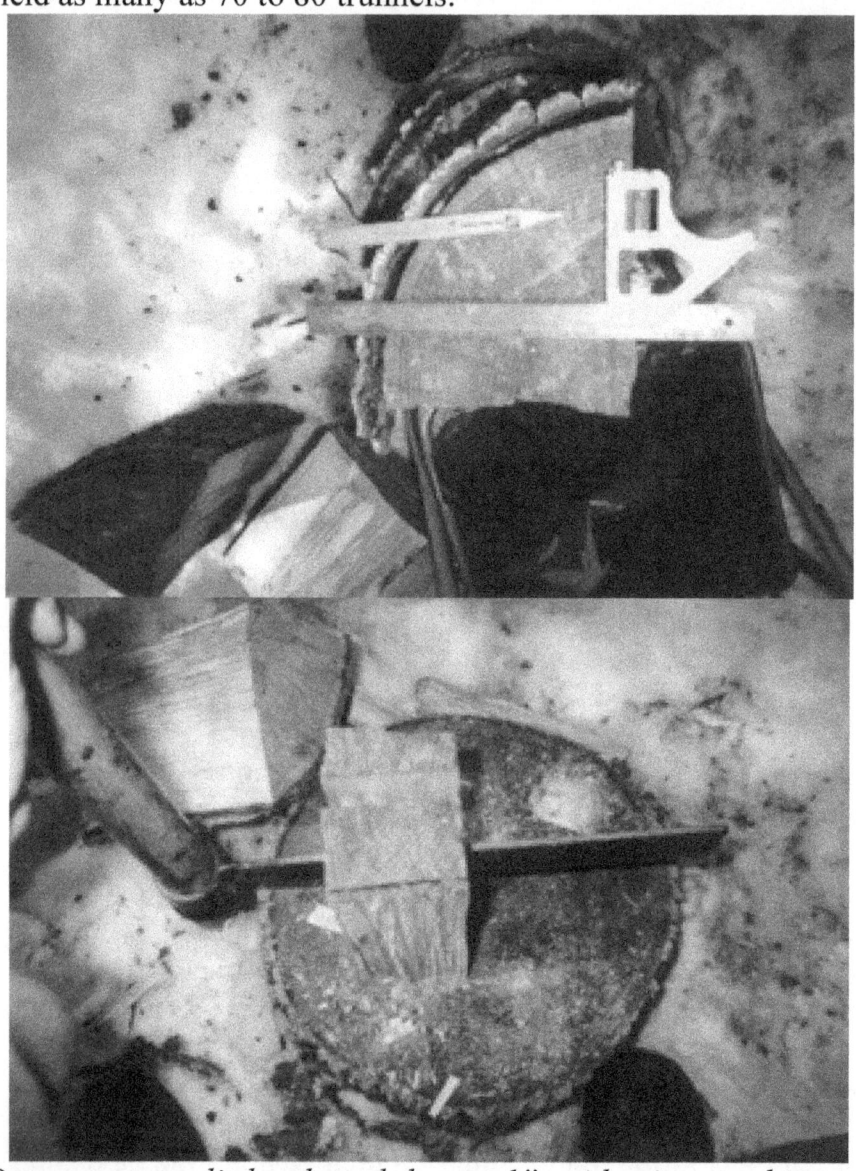

Onto quarter split hardwood, layout 1" grid using combination square (for 1" diameter trunnels). Split along grid lines with froe.

Once split with the froe, I take a hatchet and quickly remove the harsh, split corners. To finish I place the trunnel into the shave horse and, with repeated turning, shave into roughly a 1" round. I like to shave the tip of the peg to slightly less than 1" and the butt slightly greater than 1". This creates a tapered peg that will be easy to start but will have good holding ability. I finish by pointing the tip with the drawknife.

Shave round and smooth on shavehorse,

Frequently I have made trunnels without the use of the shaving horse. Instead I use a Pacific coast Native American style crooked knife.

Shaving trunnels with crooked knife.

The crooked knife eliminates the need for a device to clamp or hold the piece being worked. This is a West Coast Indian knife. This particular style of crooked knife cuts on either the pull or the push stroke.

A frame can take hundreds of trunnels. Depending upon the size of the frame, one to two days may have to be set aside solely for trunnel making.

Many framers use an additional step in trunnel making. Whereas I prefer a tapered peg, one that will start easy but will seat tight as it is driven, others prefer straight, uniform pegs. These uniform pegs are roughly shaped on the shave horse and given a point. The peg is than driven through a metal sizer. The best sizer I have seen was manufactured by Lie Nielsen for Bud Menard. He had three holes made into the metal plate: 1 1/8", 1 1/16", and 1". The shaved peg was driven through each of the three holes in succession. The result was a uniform one inch trunnel that will fit into a 1" auger hole.

Peter Campbell drives a trunnel attaching a principal rafter to tie beam.

In timber framing we want a fairly tight fitting peg. Ideally the pegs, being driven into green timbers, are seasoned. There is, however, a limit to what I mean by tight. The peg to fit should have to be driven in with a mallet, not simply pushed in. However, when the peg is too large, not even the mallet can drive it. The peg is likely to get stuck well before penetrating deep enough. With continued pounding there is a risk the peg will shatter or the tenon crack. Careful sizing of the pegs in advance prevents this. When raising a frame, I frequently prefer to assign the job of pegging to one or two people. I give them instructions on checking fit and make sure each has a knife to shave the pegs if necessary. This has proven to be both efficient and effective at raisings.

Pegs are also used in full scribe Scandinavian log building. The peg is called a Domlinger. Domlingers are used to steady the log walls and they are about 4 cm in diameter. Domlingers are placed into the logs every 300 cm. Each domlinger passes through one log and halfway into the log beneath it. This is done as every course is raised. Unlike the trunnel, the domlinger is not driven in with a mallet, but rather pushed in by hand. Because log walls settle over time, the domlinger has to be loose enough not too restrict the settling of the walls. Its real only function is too prevent the logs from tipping forward or backward.

While I was attending the Norsk Laftskoale, the domlingers we used were made of the same wood as the spruce log walls. In fact it was log cut offs that we used to make the domlingers. It was the first time I had seen softwood pegs used.

We simply split the domlinger out of the spruce waste, shaped it with the lofting axe, and pushed it in. We cut the domlingers about 2" shorter than the bored hole. In this way, when the walls settled, the peg would not lift the log above it.

I do not know if shaping domlingers with nothing more than an axe is common practice in Norway, but I have taken apart old frames in Maine in which the trunnels were clearly shaped by axe alone.

And if one gets good at splitting and shaping shingles and trunnels, what is to stop one from splitting and shaping tool handles.

Raising Day

Timber framing in particular, and log building to some extent, has this magic event called raising day. Often the day starts with neatly stacked and labeled timbers and ends with a complete frame. Sometimes the raising takes two days. In some cases only the principal timbers are raised and a smaller crew returns to put in the secondary timbers such as joists or purlins. The success of the raising depends upon how much care and preparation preceded it.

One has to be realistic about what one wants to accomplish and how it will be accomplished. Will the frame be raised by hand or will a crane be employed? How much should be pre-assembled? Do we have sufficient straps, tag lines, come-a-longs, staging, etc.? How many people will be needed and what sort of skills should they have? There are instances where maybe we do not want many people. Many years ago I put a log addition onto my log home. I had a small raising party to help me get the logs walls up, but I did not want to schedule another to help me put the knee walls, plates, ridge beam and rafters up. Simply two of us put the posts, plates and ridge beam up.

Setting plates, ridge and rafters.

I braced the king posts and put up some temporary planking. I was then able, working alone, to lift, pull, pivot and slide the sixteen foot 4"x 6" rafters into place.

Even here, however, thinking the process through was essential. The roof had a shallow 4:12 pitch. The roof rafters rested directly upon a ridge beam. If I had employed a truss system instead of a ridge beam, I could not have done it alone.

Small frames are ideally suited to raising by hand.

Many people have a fascination with hand-raising. I enjoy raising a frame by hand but one must always consider the weight of the members and the size and height of the frame.

The larger and taller the frame, the more difficult and complicated it becomes. When considering hand-raising, a careful choice of wood species can help. Spruce dries out relatively fast after it is sawn or hewn. Once partially dry it is quite light relative to other woods. Hemlock on the other hand is a very heavy wood, especially when green.

On Jim's barn we raised the bents, girts and joists by hand. A crane and a much smaller workforce were employed to place the tie beams, plates, rafters and purlins. To attempt to raise these rafter pairs on a full two story frame would have been difficult, complicated and dangerous.

Bents are pre-assembled units that are tipped up into place. In some instances it may be feasible to hand raise the bents, but it may be necessary to use a crane to raise the top plates and rafter pairs.

The first frame raising I was ever involved with was in 1980. It was a 25' x36' barn that Bill Behrens and I hewn and joined. It was successfully raised by hand thanks to Bill's planning, coordinating and skill.

The piers were leveled, the sills joined and the joists set. The deck was planked, creating a flat and level surface to assemble the bents.

All of the other timbers were cut, stacked, labeled and ready to be joined. Plenty of trunnels were made. Boards for temporary bracing and 2nd floor decking were stacked.

Bill sent out a major invite making this the big community party of the year. Many invited were carpenters by trade, but many were not. It is always wise for one to invite more people to a raising than are actually needed. About ¼ of the people who say they are coming will not make it. Of those who come, some will only be able to help for a couple of hours. Some come only to be spectators. Bill had no problem with enough people to raise the frame. There was a large potluck lunch in which Bill provided freshly smoked ham. And there was a keg of beer and a party planned for the evening. People wanted to stay.

Hand Raising Bill Behren's barn:

The piers were leveled, the sills joined and the joists set in. The deck was planked.

Floor deck created flat workspace to scribe fit the bents.

As many as 12 people lined up to lift the fairly dry Spruce bents. The braces and post provided enough hand holds so no pike poles were necessary. At the base of each post was an additional person to guide the post tenon into the mortise. Two people stood back holding ropes that had been lashed to the post tops. Their job was to pull back on the rope should the bent be pushed too far and start falling forward.

At a certain point as the bent is being raised it becomes very heavy and feels unwieldy. As it goes beyond that point in its arch the weight rapidly dissipates as the posts absorb the weight.

Once the bent was lifted and plumbed it was temporarily braced. If anything did not fit correctly, it was immediately taken care of.

Second floor joists for bays 1 and 3 were passed up and spiked into place. Decking was sent up next and the floor put down. Decking provided safe footing for setting of the plates and rafters.

To pass up the plate a rope was tied to one end. As people below lifted, Bill pulled on the rope. To help lift it high enough, one used a pike pole. As soon as it cleared the tie beam, people above grabbed and slid it onto the deck. The rope was then used to help slide the plate across open bay 2. It was then set into place and pegged. The remaining plate piece spanned only one bay and was easily lifted and placed.

Scaffolding was assembled to make raising the rafters easy. The pole rafters were passed up and assembled on the deck. Only every other set had a collar tie. Once assembled the

rafter pair was tipped into place, plumbed and temporarily braced with a board spanning several rafters. Another person bored the hole for the trunnel into the plate and pegged it.

To allow room to tip up the end rafters, the last 7 pairs were tipped up tight one against the other. Once all were assembled, these last few were slid along the plate to where they were to be pegged.

Just as the sun was setting the frame was finished.

The barn 29 years later.

Bill had used a lightweight, common rafter system with simple joinery that made raising by hand manageable. Bill has since sold the property. The barn with its clapboard front and shingled sides sitting next to an old cape gives the appearance of an old English barn. Some who see it think it is as old as the 170 year old house, but a look inside reveals joinery of a different era.

In raising a frame there is a definite sequence in how the timbers go together. When I am involved in designing a frame, I always consider how the frame is to be raised. I mentally raise it to make sure that I fully understand the sequence in which the frame needs to be raised. Straying from that sequence can easily create a situation whereby one must disassemble some of the assembled frame to fit in a piece that needed to be done at a specific moment in the raising. A tenoned brace is a good example. The brace must be fitted at the same time the girt or plate is fitted to the post.

Mortise and tenon braces must be set as the frame is raised.

In some ways that was some of the magic of Bill's frame. Bill, either by chance or plan, managed to avoid all of the difficult aspects of raising a frame. The braces that run from plate to post were scribe-fitted with exterior laps after the raising. All the people had to contend with was setting the plate. There was no need for people to be balancing and trying to fit in braces as the plate was being set down. There also were no connecting girts tying the bents together with mortise and tenon joinery.

Many of the structures I have worked with of late do not conform easily to hand-raising. The structures are either too large, the manpower insufficient for hand-raising, or components of the structure are most easily assembled upon the ground and lifted by crane into place.

Raising Pre-Assembled Roof Trusses with Crane:

Trusses assembled on ground. Strap and tagline attached and truss lifted.

Tagline steadies truss from rotating as it is floated over. Two people catch rafter tails and ease into place.

First truss up and braced. *All six trusses up.*

Every frame has a sequence that must be followed in raising. The sequence will vary depending upon the frame design. In some, the post must be raised individually and the plates, girts and tie beams placed one at a time. On other frames, the eave walls are assembled and raised as units, with the girding and tie beams set individually. The most common frame design requires that bents are pre-assembled and raised. Connecting girts are fitted between the bents and the plates are set atop the bents.

Raising a Conventional Timber Frame with a Crane :

Sills, sill girders and joists set.   *Bents pre-assembled.*

Bents are floated into position.

Left: Girts are set between bents as bents are raised.
Right: 2nd floor joists are dropped in.

Plate scarf joint assembled and stabilized for raising with crane.

*Plate set along with **post to plate** braces, followed by rafter trusses.*

Purlins dropped in.

The finished frame

Regardless of whether the frame is raised by hand or raised with a crane, if the resources are available and the sequence has been carefully thought through, the raising will be successful.

Cruck being eased down between plates.

Building with logs

Continental Europe traditionally made little distinction in name between timber frames and log buildings. Log buildings were called full-timber buildings while timber frames were called half-timber buildings. Purists in the log building craft, however, accept a more rigid definition. Just as the timber framer tries to differentiate his craft from simpler versions of post and beam construction, log builders attempt to differentiate their craft from simply post and beam construction with log infill. Unfortunately, I am not a purist.

Log building in Europe had reached a high level of sophistication and skill centuries ago. Supposedly log construction was developed and refined by the Russian people. From there it spread westward through Europe. Each locality developed changes in construction techniques as the different styles of building became culturally entrenched. The large Bauernhaus of the Alps is radically different from a simple Scandinavian dwelling or storage house. Whereas the Alpine builder would attempt to put a very large structure beneath a broad chalet roof, the Norwegian would be more apt to build several small structures.[8]

Log farm buildings (Open Air Museum, Lillehammer)

In North America, highly skilled, weather tight log building is a fairly recent development, mostly influenced by the Scandinavian method of coping the logs to fit one onto the other.

Here in northern New England local histories describe a plethora of log buildings being built by pioneer settlers, yet almost none of them remain. The log shanties and huts were abandoned as soon as they could be replaced by timber framed buildings. They were viewed only as temporary dwellings.

Despite the fact that these buildings were temporary, were built with few resources and a great deal of haste, they have become romanticized. When in Norway, I explained how in North America we chink between the logs, and how sometimes homeowners want the chinking space to be as great as 4" wide. This baffled them. In Norway chinkless construction is the goal. One should be able to create a weather tight home without

---

[8] Phleps, Herman, *The Craft of Log Building,* Lee Valley Tools, Ltd.: Ottawa, Ontario, 1982.

relying on synthetic or masonry chinking. They saw no advantage to log construction that needs chinking.

In 2005 Bill Rispoli and I built an unusual structure that many would say was not a log building at all but a timber frame with log infill. The Quebecois would have called this style of building *piece sur piece avec coulisses.*[9]

Piece sur piece avec coulisses.

The logs for the walls were hewn on two sides only keeping the top and bottom round. The ends of the log were tenoned. A groove was cut into the post with a chainsaw, mallet and chisel. The logs were then slid down between the posts with the tenon sliding in the groove.

Because, Myles, the owner, wanted a large 4" chinking space, blocking had to be inserted and screwed into the groove after every log was placed. The log was then pegged to the post with two small trunnels.

---

[9] Gauthier Larouche, Georges, *Evolution de la maison rurale traditonelle dans la region de Quebec*, Les Presses De L'Universite Laval : Quebec 1974, p. 84

Once the structure was built there were some delays that postponed the chinking by over a year. The delays proved beneficial. The logs were given time to season and shrink. I have not met the person who did the chinking, but to chink such a wide opening with synthetic chinking is exceptional. By all appearances the chinking has held up well.

The cabin prior to raising the rafters

I seldom build with such large chinking spaces. I was hesitant when I built this structure. I tried to convince the owner to drop the chinking space to two inches. He, however, knew the look he was going for. To see the cabin now, I realize he was right.

Living in New England, I am not often approached to build with logs. New England is timber frame country. When I have built with corner-joined logs, invariably I have used half-dovetail joinery.

There is a strong precedence for hewn log construction employing half or full-dovetail corners. Though few examples of log construction remain in the East, of those that do remain the majority are hewn and joined with either of the two versions of dovetail joinery.

Hewn log walls joined with half-dovetail joinery

Checking out a half-dovetailed log house on the Blue Ridge Parkway, Virginia.

From Quebec to Georgia examples of this form of construction can be found. The oldest British fort in the US still standing is Fort Western in Augusta, Maine, built in the 1760s. To see the building on the outside, one easily imagines one is looking at a long timber framed building covered with shingles.

Beneath the shingles lie hewn logs, the corners joined with half-dovetail joinery. In the Northeast, covering the log structures with shingles or clapboards appears to have been the norm, making examples of this form of construction almost invisible. Down in the Appalachian South examples of hewn log construction are easier to find because they are not necessarily covered with either shingles or clapboards. With winter coming later and leaving sooner then their counterparts experience in the North, these cabin dwellers may not have felt the need as strongly to weather tight the cabin.

Just as in New England the revival of timber framing was being pioneered by people in the early 70s, Peter Gott and Charles McRaven brought about the revival of hewn log construction in the southern Appalachians.

Gott, in particular, developed a mathematical formula for the layout of half-dovetail joinery. Using a chalkline he would snap a centerline down the face of the timber. He would plug numbers into the formula below and, using the answer received and the centerline, he would lay out the half-dovetail joint.[10]

$$\text{Depth of cutout} = \frac{\text{height of log} + \text{chinking thickness}}{4} - \frac{\text{rise of notch cutouts}}{2}$$

I used to slope my dovetails with a 2" rise but of late I have reduced the rise to 1 ½". If we use a 10" diameter log with a chinking space between the logs of 1" and the rise of the notch is 1 ½", we can plug the numbers into the formula and find what the depth of the cutouts should be.

$$\frac{10" + 1"}{4} - \frac{1 \text{ ½}"}{2} = 2"$$

---

[10] Langsner, Drew, *A Logbuilder's Handbook*, p.130.
Petersen, David, *Building the Traditional Hewn-Log Home,* Mother Earth News, July/August, issue #94, Mother Earth News.com.

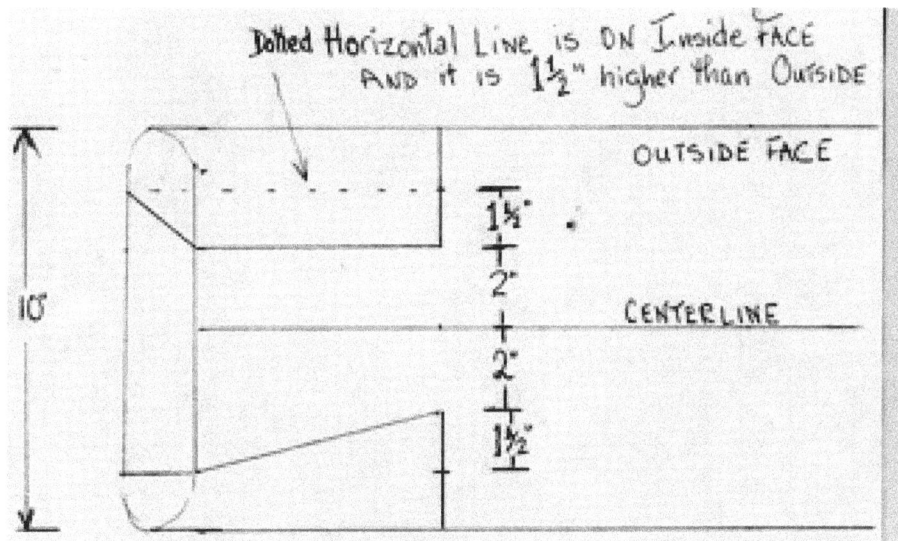

Half-dovetail layout for 10"x6" hewn log with 1" chinking space and a dovetail slope with a 1 ½" rise.

I have cut half-dovetais using a handsaw, chisel and adze. I have also cut them using a chainsaw and axe. My personal experience has led me to believe the chainsaw is by far the most practical tool to use when cutting half-dovetails.

Cutting half-dovetail log notches:

Underside notch laid out.

Cutting underside notch with chainsaw.

Cutting back edge of top notch

Cutting along slope line of top notch.

Notches are cleaned up with chisel.

As long as there is a chinking space, Gott's formula works well. Hewn log building in the southern Appalachian mountains generally had a chinking space of from 2" to 4" in height. Gott's formula brings log to within a variance (plus or minus) of that chinking space. The difficulty occurs when the chinking space is 0". Logs may have crown or humps and bumps, hewn surfaces may be irregular, and the plus or minus variance may also make it difficult to seat the log along its whole length and at the joint.

When the chinking space is zero, I find it easier to use Gott's formula for the top notch only. For the bottom notch I use a combination square and calculate the depth the notch must be cut directly off of the preceeding log the joint must fit. Even with careful calculations, I often find I must re-scribe and re-cut the underside notch if the cut was too shallow or, if the underside notch was cut too deep, I must plane the top surface of the log upon which this log must sit.

Half-dovetail joints with no chinking space between sawn logs above and hewn logs following page.

When I leave no chinking space I either have the top and bottom surfaces of the log sawn or hewn flat. Chinkless, however, does not mean weather tight. To achieve that the logs would have to be scribed using a method such as the Scandinavian cope.

When logs are laid flat on flat, they still need to be sealed, usually on the outside with a small bead of synthetic chinking spread over small diameter backer rod pressed into or up against the crack between the logs.

I have seen and used sill seal as a material to lay between each course of logs as they are laid up. I, however, would not use it now. If the synthetic chinking should fail and the underlying backer rod beneath the chinking not provide a watertight seal, the sill seal itself may not make enough of a watertight seal in a wind-driven rain.

What I lay now upon the logs as the courses are being placed is acrylic-modified, asphalt-impregnated sealant tape. This tape usually comes in half-inch width, is compressed to 1/8" thickness and springs upward to almost 5/8". The tape adheres to the log below, and as the next log is placed upon it, it expands to fill any voids and adheres to the log above creating a watertight seal.

With sealant tape, if one wants to wait and allow the logs to season before applying the backer rod and synthetic chinking, one can. Two strips of sealant tape should keep the structure watertight until it can be chinked the following year.

Sill seal and backer rod are generally available at most building supply centers. To purchase synthetic chinking or sealant tape one generally must make contact with a log home supply company. There are several on-line marketers of materials for log builders. Anyone considering building with logs should make contact with one of these marketers and have a catalog sent. There are materials, tools and supplies that are available nowhere else.

Sealant tape should not be necessary. Synthetic chinking, if properly put on, should make a strong weather tight seal. Unfortunately, if not properly applied, it can fail. The sealant tape will protect against leaks even when the chinking fails.

When I was in Norway, attending the Norsk Laftskoale, I learned how to scribe one log to fit upon the other. Using chainsaw and a channel knife I would create what most log builders call a Scandinavian cope. (See section *Tools)*

One competent in scribing and cutting the Scandinavian cope could also scribe the half-dovetail at the same time, creating a log structure that is chinkless, weather tight, and joined using half-dovetail joinery. This has been done in Estonia as well as here in the States. Robert Chambers, a highly skilled log builder, built just such a house using scribe-fit logs and full dovetail joinery. A complete description is in *Fine Homebuilding,* August/September 1986, No. 34.

Log buildings have other characteristics unique to them: settling and shrinking. Wood shrinks in both width and thickness, but not in length. In timber framing this is not considered a major problem. Timber framers continue to work with green timbers. Timber frame restorers continue to marry 200 year old timbers to fresh cut timbers. Certain techniques are used to minimize the impact of this shrinkage, but little else is done about it.

In log construction it is of major concern. The Log Builders Association has adopted the standard that ¾" shrinkage allowance must be made for each foot of wall height. This implies that in certain log building styles an 8' tall wall of logs can lose 6" in height.

A building losing six inches of height in less than 3 years can experience some very destructive reactions. Rather than the log walls absorbing the weight of the roof, the weight of the roof may instead become absorbed by window and doorjambs causing them to bind, jamb and cease to function properly. The staircase to the second floor will push outward and the treads will no longer sit level. Plaster walls will crack. Even PVC drainpipe from the second floor can crack under the pressure.

Log builders must plan for shrinkage and settling. Despite this fact, Gott and McRaven never seemed too concerned with settling. McRaven simply says to raise the log walls then wait for them to settle and shrink before nailing the 2" thick jambs directly to log ends for window and door openings. It is doubtful he waits 3 years before nailing up jambs, so something else is obviously going on.[11] In Scandinavian scribe-fit log construction, the outer edges of the cope compress into the log beneath. Compression space is even built into the corner locking-joint. The result is increased settling. The settling is actually encouraged in this system because as the logs settle the structure becomes increasingly weather tight.

If we contrast this with hewn half-dovetail log construction, we see strong dissimilarities. In half-dovetail log construction a chinking space of from 2" to 4" is the norm. The logs do not touch each other except at the corners. The half-dovetail joint is a flat joint. The ability of one log to compress into another is minimal.

---

[11] McRaven, Charles, *Building and Restoring the Hewn Log House,* p. 80, 81.

Maximum shrinkage occurs in the sapwood. At the half-dovetail joint, the sapwood is removed.

And there is one other factor that may play into the reason why hewn log structures appear to have less problems with shrinking and settling. It takes an awful lot of time to hew enough timbers for a log house. By the time the last log is hewn, a good portion of the logs may have already done the bulk of their shrinking.

Unless the logs are well seasoned and a chinking space is provided for, I recommend allowing for shrinkage and settling.

Wherever there are vertical members, window and doorjambs, posts, etc. allowance must be made for settling. Two by fours are the common material used. These are fitted into a channel cut into the log ends, allowing logs to settle around verticals, but the Code of the Log Builders Association also allows one to use angle iron. Whether two by fours or angle iron are used, the tool to cut the channel into the log ends is the chainsaw. If using angle iron, this is easy. One cuts the door opening to the width needed, adding the extra for the jambs, then one snaps a chalkline down the middle of the log ends and using the chainsaw, one plunge cuts to the desired depth and then just follows along the chalkline. A single cut suffices for angle iron, where if the two by four is used, one must, using the chainsaw, widen the groove to 1 ½".

Door opening grooved for two by four.

In the proceeding Norwegian log building, door opening is cut a good 6" taller than necessary. A two by four is inserted into groove cut into logs. Jamb is nailed to two by four. Logs, as they settle, slide along two by four. Extra 6" of space above insures log walls will never compress doorjamb.

Left – Groove for angle iron cut into door opening.
Right – Angle iron test fitted before attaching iron to door jamb.

Window jambs attached to angle iron. Space above jamb is allowance for settling as logs slide down along angle iron.

Because the posts heights will remain constant as the log walls settle, a means to lower overall height of posts needs to be incorporated. Screw jacks set atop the post, hidden beneath removable trim, enables the owner to periodically lower the height as the logs settle.

Using electric planer, a 3/8" relief was created to keep back of angle iron flush with post. Angle iron was then centered and attached to post with screws.

Post set into log wall.

Windows and doors of this home have settling allowance. Walls are chinked using backer rod and synthetic chinking. And roof has ample overhang. If screw jacks are properly lowered in first

three years and chinking was correctly applied and periodically checked, this building should provide a comfortable home for years to come.

A log home interior.

In the End

In Walden, Thoreau lamented the fact that people forever surrender the joy of building their own dwelling to the carpenter. We allow it all to seem too complicated, requiring altogether too much skill and hire it out to others, when in reality it need not be that complicated.

In this day and age, if one wants to build, there should be little reason to stop one. As for timber framing and log building, there are books on the subject covering every aspect from tree selection to interior finishing. There are log building and timber framing schools offering both introductory building courses as well as complex and advanced building courses. And there are opportunities to learn by volunteering. The above structure was built for the Maine Organic Farmers and Gardeners Fairgrounds entirely by volunteer labor. One volunteer, Peter Campbell, worked faithfully one day a week on the project until it was completed. In exchange, he learned how to hew timbers and cut joinery.

If one does not want to go it alone, approach a builder who is willing to allow you to work with him/her on the project. On every project I have taken on, the owner has been invited to work on the project as little or as much as he/she chose. I have had some builder-owners who have taken to the project so wholeheartedly that they have done more of the work than I did.

Jim Lattin hanging the pine bough on his 24'x 28' two story barn. Jim worked tirelessly on his barn frame, putting in far more hours than I did.

Jim Farrel working on the log/frame addition to his house. Jim clearly matched me hour for hour on this frame, and then finished the addition alone, roofing, sheathing, insulating, etc.

And then one can do what my brother Den did close to thirty years ago. Working with one book on timber framing, no experience, no electricity, a very small chainsaw, a couple of axes and a few hand tools, he walked into the woods and built his hewn timber frame.

Den (back row, third from left), standing proudly upon his newly raised frame.

That, after all, is what Thoreau did. If one is contemplating building a timber frame and is experiencing some self-doubt, read his account: borrowing tools, felling trees, hewing, cutting joinery, plastering walls, and building a fireplace. It is improbable one could stay within his $50 budget, but any and all labor supplied by oneself is labor that one does not have to hire, and any and all building materials created by oneself is material that one does not have to purchase.

Will the joinery be flawless? Doubtful. Will mistakes be made? Probably. Flawless joinery is craftsmanship learned. Some learn it quickly; others take quite awhile. I am a slow learner, but, with every structure I build, the craftsmanship improves. The more timber layout one does, the more one uses the tools, the higher the quality becomes. And mistakes are a part of life. It is how we learn. A mistake made on one frame is a lesson learned. It is doubtful the same mistake will be repeated on the next frame.

Not one of the early frames I built has collapsed, even though not one had flawless joinery. I have yet to witness a frame built in which at least one mistake was not made. Each frame one builds adds to ones field of experience and level of skill. The way to learn is to do.

The joy and satisfaction that one can experience building is immense. No occupation I have had has been as profoundly fulfilling as crafting frames of timber, and few occupations reward one as much as the feeling of accomplishment one gets seeing a finished frame standing tall and new.

The raising itself is a social gathering, it is a joyous celebration with friends.

Raising a frame with friends, followed by drink and food is a social gathering that is hard to beat.

And if the frame is for oneself, one gets to decide how one will live in it. There are various options for enclosing a structure and various ways to insulate a structure. There are windows and doors to choose. There are diverse ways to finish and trim both the interior and exterior. There are natural oil finishes today that have no petroleum distillates and natural clay plasters to apply to wallboard, giving walls a wattle and daub feel and look. The options are only limited by our imagination.

Further Reading

Beaudry, Michael. *The Axe Wielder's Handbook.* Horizon Publishers: Springville, UT. 2005

Chambers, Robert. *Log Construction Manual.* Deep Stream Press: River Falls, Wisconsin. 2003

Chambers, Robert. "Scribe-Fitting a Log House." *Fine Homebuilding.* August/September, No. 34. 1986

Chappell, Steve. *A Timber Framer's Workshop.* Fox Maple Press: West Brownfield, Maine. 1998

Elliot, Stewart and Wallas, Eugene. *The Timber Framing Book.* Housesmiths Press: York, Maine. 1997

Gauthier-Larouche, Georges. *Evolution de la maison rurale traditionelle dans la region de Quebec.* Les Presses De L'Universite Laval Quebec. 1974

Harris, Richard. *Discovering Timber-Framed Buildings.* Shire Publications, LTD. 2001

Heavrin, Charles. *The Axe and Man.* Astragal Press: Mendham, NJ. 1998

Hunt, W. Ben. *How to Build and Furnish A Log Cabin.* Collier Books: NY. 1974

Kahn, Lloyd. *Builders of the Pacific Coast.* Bolinas, CA. 2008

Kahn, Lloyd. *Home Work: Handmade Shelter.* Bolinas, CA. 2004

Kahn, Lloyd. *Shelter.* Shelter Publications: Bolinas, CA. 1973

Langsner, Drew. *The Logbuiler's Handbook.* Rodale Press: Emmaus, PA. 1982

Mackie, Allan B. *Building with Logs.* Firefly Books: Buffalo, NY. 1997

McRaven, Charles. *Building and Restoring the Hewn Log House.* Betterway Books: Cincinatti, OH. 1994

Newman, Rupert. *Oak-Framed Buildings.* Guild of Master Craftsman Publications: East Sussex. 2005

Petesen, David. "Building the Traditional Hewn-Log Home: A Mini Manual", *Mother Earth News,* Issue #94, July/August 1985. www.motherearthnews.com/library

Phleps, Hermann. *The Craft of Log Building.* Lee Valley Tools, ltd.: Ottawa, Ontario. 1982

Sobon, Jack A. *Historic American Timber Joinery: A Graphic Guide.* Timber Framers Guild: Becket, MA. 2002

Timber-Framed Houses. The Taunton Press: Newtown, CT. 1992

Thoreau, Henry David. *Walden*

Wilbur, C. Keith. *Homebuilding and Woodworking in Colonial America.* The Globe Pequot Press: Old Saybrook, CT. 1992

Index

Aspen, 36-37, 47

Ax, Hewing, 12, 35-36,
 39, 41-43, **79-80**
 Lofting, 92

Brace, **129-142**
 Compression, 130
 Mortise & tenon
  129-138
 Scribed, 138-142
 Square rule, 135-138
 Tension, 140-141
 Wind, 165-166

Cruck, 20-21, 26-30,
  195-203
 Layout grid, 198
 Layout and cutting,
  198-201
 Model of, 21
 Spur, 195-196

Dovetail, half
 Brace, 141-142
 Log building,
  242-249
 Shouldered, 117

Dovetail, full, 149
 Joists, 124-127
 Purlin, 187
 Shouldered, 118,124

Girding beam, 158-159

Girt, 234
 Chimney, 112, 117

Hewing, 12, 35-43, **51-82**
 Joggling, 63-65
 Log dogs, 57
 Scoring, 59-63

Joists, **120-128**
 Dovetailed, 124-127
 Hanging, 127
 Locust, 22-23
 Mortise, 120-121
 Tenoned, 123

King post,
 Braced, 145, 236
 Strut, 191
 Truss, 189-191
  243

Mortise, **99-101**, 107
 Blind, 113-114
 Brace, 133-138
 Bridled,116, 170,
  172
 Joist, 120-121
 Square rule. 107

Log Building, **249-266**
 Half-dovetail,

38, 252-257
Locking joint, 92
Piece sur piece, 251-252
Purlin, 161
Scribe-fit, 93-95
Sealant tape, 259-260
Settling allowance, 261-265
Sills, 120-121

Pine, 34-43, 47, 77
Shingles, 223-229

Plate, 24, 97-99, 107, 130-131, 143, 147-158, 164-167, 173-182, 195, 202, 235, 237, 240-247

Post, Jowled, 151-152

Purlin, 20, 43, 158, 161, 163, **185-193,** 237, 246

Queen post, 145
Truss, 191-193

Rafters, 20, 28, 43, 97, 143-146, 154-155, 160-161,**163-193,** 195-197, 202, 246 235-236, 240-241
Beveled stop, 178
Birdsmouth, 97, **173-177,** 181-182
External, 173-176, 181
Internal, 176-177
Beveled stop, 182
Calculating, 167-172,182
Collar tie, 109, 189
Common, 163-168
Level cut, 169
Plumb cut, 169
Post connection, 183-185
Step-lap, 177-182
Tenoned, 154-155, 182

Raising frame,
Crane, 243-247
Hand, 236-243

Reference face, 91, 99, 102, 106, 108, 132-136, 152, 154, 206-207

Ridge beam, 145, 236
Board, 166

Roof pitch, 164-166

Scarf joints, **205-221**
Bladed, 212-216
Halved and bridled, 211-212
Scissor, 208-210

Stop-Splayed,
 206-208
Splayed and Tabled
 216-221

Scribe rule, 108-109
 Brace, 138-142
 Cruck, 198-202
 Log, 93-95

Shakes, Shingles, **223-229**

Sill, **111-120**
 Bridled, 116
 Girder, 114-116
 Lap joined, 116
 Locust, 22-23
 Log, 119

Square rule, 105-108
 Brace, 135-138

Summer beam, 116

Tenon, 99, **101-105**,
 107-108, 190-191,
 211, 214
 Brace, 132-133, 139
 Joist, 123
 Sills, 113-116
 Teasel, 151, 153

Tie beam, **143-161**,
 182-183, 189-190,
 195-196, 201-203

Tools, **81-93**
 Circular Saw, 87-88
 Chain mortiser,
  85-86
 Channel knife, 94
 Crooked knife, 231
 Drawknife, 228-229,
  231
 Framing chisels, 88-89
 Froe and club, 224-227
  230-231
 Mallet, 90
 Self-feed bit, 84
 Slick, 90
 Speed square, 171
 Shave horse,
  228, 231

Trunnels, **229-234**
 Domlinger, 233-234

Truss,
 Hammer beam, 160
 King post, 189-191,
  243
 Queen post, 191-193

Tying Joint
 English, 149-156
 Half-dovetailed,
  147-149
 Through-tenoned,
  146-147

Wood, characteristics,
 45-48, 77-80

www.ingramcontent.com/pod-product-compliance
Lightning Source LLC
Chambersburg PA
CBHW031137160426
43193CB00008B/169